Telecommunications
Networks

ℍ
HOWARD W. SAMS & COMPANY/HAYDEN BOOKS

Related Titles

Digital Communications
Thomas C. Bartee, Editor-in-Chief

Data Communications, Networks, and Systems
Thomas C. Bartee, Editor-in-Chief

Micro-Mainframe Connection
Thomas William Madron

Computer Connection Mysteries Solved
Graham Wideman

Modem Connections Bible
Carolyn Curtis and Daniel L. Majhor, The Waite Group

Printer Connections Bible
Kim Marble and Jeff House, The Waite Group

Optical Fiber Transmission
E. E. Basch, Editor-in-Chief

Turbo C Programming for the IBM
Robert Lafore, The Waite Group

Microsoft® C Programming for the IBM®
Robert Lafore, The Waite Group

C Primer Plus, Revised Edition
Mitchell Waite, Stephen Prata, and Donald Martin, The Waite Group

Advanced C Primer + +
Stephen Prata, The Waite Group

Topics in C Programming
Stephen G. Kochan and Patrick H. Wood

Programming in C
Stephen G. Kochan

C with Excellence
Henry F. Ledgard

For the retailer nearest you, or to order directly from the publisher,
call 800-428-SAMS. In Indiana, Alaska, and Hawaii call 317-298-5699.

Telecommunications Networks
A Technical Introduction

R. J. "Bert" Murphy

HOWARD W. SAMS & COMPANY

A Division of Macmillan, Inc.
4300 West 62nd Street
Indianapolis, Indiana 46268 USA

©1987 by R. J. "Bert" Murphy

FIRST EDITION
FIRST PRINTING—1987

International Standard Book Number: 0-672-22588-3
Library of Congress Catalog Card Number: 87-61013
Acquisitions Editor: *James S. Hill*
Editor: *Tom Whipple*
Interior Design: *T. R. Emrick*
Illustrator: *Don Clemons*
Indexer: *Northwind Editorial Services*
Cover Graphic: *Meridian Design Studio Inc.*
Compositor: *Shepard Poorman Communication Corp., Indianapolis*

Printed in the United States of America

All photographs are reproduced with the express permission of AT&T Bell Laboratories, Naperville, Illinois.

Trademark Acknowledgements

All terms mentioned in this book that are known to be trademarks or service marks are listed below. In addition, terms suspected of being trademarks or service marks have been appropriately capitalized. Howard W. Sams & Company cannot attest to the accuracy of this information. Use of a term in this book should not be regarded as affecting the validity of any trademark or service mark.

IBM and IBM-PC are registered trademarks of International Business Machines Corporation.
Picturephone is a registered trademark of AT&T Technologies.
DECNET is a registered trademark of Xerox Corporation.
Telenet is a registered trademark of Telenet Communications Corp.

Contents

Preface

A new phenomenon, telecommunications networking, has burst on the scene during the past few years. This hybrid of a half dozen fields of study has led to the need for a complete understanding of communications and computers, including the various aspects of voice and data transmission, and for individuals who perceive how these units operate and function within a network. We need individuals capable of introducing changes as they arise.

This book is about telecommunications networking and where we will be led by it. Although it deals with historical and technical points, its main purpose is to give insight and knowledge to those individuals needing a deeper understanding of this new phenomenon.

The first part of the book gives an introduction and a history in order to reach a definition of this new "information age." There is no apology for the use of history to obtain this definition, for the very process reveals a uniqueness of telecommunications within this age. There exists a need today to combine computers and communications into a new science much the way Descartes combined algebra and geometry into a new science called analytic geometry—a science which greatly expanded man's knowledge. Perhaps the same historical insight is necessary to combine computers and communications and, thereby, realize the same growth.

The book then discriminates among different views of telecommunication—transmission, traffic, data, and others—with explanatory and illustrated material. We first define these parts so that we can enter the world of network arrangements. The book concludes with some practical examples on how these parts are put together in networks and gives a view of the future for telecommunications.

Some approach telecommunications networks as a natural evolution of their knowledge base, some from involvement with data, some as entrepreneurs seeing the opportunities, and others seek a money-savings approach for a widespread organization. Accordingly, the demands on telecommunications are as diverse as the motives leading to it. Books written to satisfy these demands are similarly diverse and have difficulty satisfying one, much less more, of these demands.

One way to overcome this difficulty is to write a book so comprehensive that all problems of telecommunications are treated in it. This is not practical because the size would be overwhelming. A few pages devoted to each problem is not the answer either; it would do little for the novice reader and insult the intelligent telecommunications person. The solution, therefore, is to select some of the major problems (or opportunities) and devote sufficient time to each without reaching unmanageable lengths.

The overall position taken in this book is that telecommunications, as an operational part of life and work, is both powerful and pervasive. Its power stems from its ubiquity and demand for instant recognition. Its pervasiveness is seen today in telemarketing and any effective TV ad campaign. The combining of these two marketing tools is starting to be seen on a limited basis as certain cable stations start to sell directly.

There is no question that today's telecommunications individual is striving desperately for understanding in this rapidly changing environment. The continued round of technological changes, the babbling about ISDN, the succession of communication companies seeking a marketing position, the unclear position of the federal government both domestically and internationally, all appear to give the impression of an unstable telecommunications environment. The current fear is that through artificial intelligence, the computer will do the thinking.

But as a longtime telecommunications person specializing in various aspects of the field, I find these several denigrations of the individual ability to handle telecommunications unconvincing. It will be our thinking and not computers or government regulations that will make future telecommunications networks. To reach this platitude, I have presented a capsule view of the current problems associated with telecommunications and why they will require a human solution.

There are innumerable terms and equations, since each of these subjects has a special language that would have to be introduced if the treatment were to be complete. However, this is not a highly technical book with vast and complicated equations. The subject material is approached from the viewpoint of "practical" learning, with the purpose of being sufficiently complete to allow the reader to define links to other aspects of telecommunications in order to "put it together."

One goal of this new information age is to create a time when knowledge is universally available and we have overcome distance, standards, and language. To reach this goal, we need a knowledge of telecommunications networks.

R. J. "Bert" Murphy

Introduction

It seems impossible to conceive of a more gigantic megamachine than the telephone network interconnecting the world through its various nodes and switches. The capability of the system—where a person can sit by a phone and dial anyone in the world—is staggering. Now picture this system, with its capability of transmitting voice, data, and image anyplace in the world by phone, replaced by a multifunction terminal/phone and you will have an idea of how the present system will evolve to an information network.

The new telecommunication and information society will pose a serious problem for the person considering a communication network because a plethora of systems can be purchased or leased; the dependency on Ma Bell appears to be waning, and people are wondering why. "Why break up something that works?" The answer lies in the range of technologies the customer will need to operate in the present and future environment. An analogy may be the successful company carrying messages across the country via pony express when someone installs a transcontinental railroad alongside the trail. As the railroad changed the transportation system of the country, the new technology will change the communication system of the world.

Business and government need to transport information other than voice (data, facsimile, image) as rapidly as possible from end to end. Thus, a new information transportation system is needed, and, hence, the network must be opened to competition to allow for more innovative ways to accommodate these demands. Otherwise, new networks will emerge, eventually replacing the current network.

The trend is toward a complete telecommunication network that will enable anyone to communicate nationwide or worldwide in any desired form: by voice, text, data, and video, separately or simultaneously. This will first come about in private telecommunication systems, because for years businesses have been a main driving force for telecommunication improvements.

Fiber optics will revolutionize the distribution (wiring) system from the switching machines to the customer and will eventually allow a video telephone at the home. In addition, the Integrated Services Digital Network (ISDN) will permit the interconnection of voice, data, text, or video from one person to the other in any location and connected to any system.

Fiber optics and ISDN, together with the expected electronic evolution, will bring about a completely new telecommunication network, regardless of deregulation or other activities in the field.

A multifunctional terminal is emerging that employs a video display telephone, an electronic telephone directory, a notepad, and a keyboard all rolled into one. The terminal is also the area where the computer industry and the telecommunication industry will clash. The computer companies now have an excellent market with their terminal, keyboard, and printer interconnected to a mainframe or other computing device. If you look at the video display telephone, you will also see the emergence of a computer terminal and keyboard.

With both the computer and telecommunication companies competing for your business, and the ability of smaller chips to do more, you have a situation where even the best science fiction writer won't be able to anticipate all the features and services available from these terminals.

The capabilities of the telecommunication network will increase enormously as these various services grow, and businesses and individuals will depend on the telecommunication network as their transportation system. This transportation system will be used to interface to the office, to do most of the shopping (although people will probably do actual buying), and to bring information/entertainment to the home, including newspapers and magazines. Obviously, all these riches will not come free, but money may not be the only consideration. The increase in productivity and the ability to work at home yet be tied to the company's network are part of the decision.

The requirements for a national standard compatible with an international standard currently loom as the greatest deterrent to achieving these goals.

Under autocratic AT&T, the long-distance network has played a special role in telecommunication, for it was the symbol of power. For 100 years, AT&T set the standards for our networks. They ruled knowledgeably, as the network, with its ability to interconnect any pair of phones, demonstrated.

In the early 1960s, Tom Carter designed a device for coupling the automobile radiophone with the telephone network to enhance the reach and quality of the phone connection. AT&T prohibited the attachment of the device to the network, and Tom Carter fought back. Thus, a battle was started, which initially culminated in the Carterfone decision and, more importantly, began a sequence of events leading to the end of the natural monopoly known as AT&T.

In the early 1970s, the U. S. Justice Department brought an antitrust suit against AT&T. They wanted to break up AT&T without impairing its valuable service. The trial was delayed year after year until it appeared that there would be no trial. Finally, Judge Greene stated that the trial would begin in 1981. Much publicity and great arguments surrounded the trial, and, for many, it became daily soap-opera reading and viewing the youthful Justice Department lawyers against the AT&T attorneys.

In an unprecedented action, the Justice Department and AT&T announced a settlement on January 8, 1982. AT&T agreed to divest itself of 22 local operating companies; in return, it would be allowed to compete in unregulated industries, such as computers.

The announcement caught everyone, including Judge Greene, by surprise, and though this signaled the start of modern telecommunication, no governmental agency was prepared to monitor this new industry. As a result, the activities of AT&T, the Bell regional companies, and GTE are controlled by Judge Greene. Senator Dole has introduced a bill to move this control to the Federal Communications Commission (FCC).

Most people were overwhelmed that Bell divested 60% of the company and broke that segment into seven regional telephone companies. The segment divested was the income-producing segment of AT&T, whereas the segment retained was the profit center of new business.

Equally important, the breakup destroyed the symbol of power for AT&T—the network. The number of networks, both private and public, has skyrocketed since the agreement and represents the biggest change so far in telecommunication.

Now, a company as big as General Motors can build a private network and save money on their telecommunication charges. This was not the intent of deregulation. Unless a small company or businessman can realize some savings and feature improvement, the experiment will be a failure. The government, to support its decision, must establish open standards for interconnecting to the network without harassment from AT&T. Unfortunately, these standards should emulate from the FCC, which is more concerned with policy. Establishing standards is hard and dirty work, but if this new telecommunications is to work it will work because of standards and nothing else.

The present framework for deregulation of telecommunications in the United States is not at all a laissez-faire structure. It reflects the desire of AT&T to enter areas formerly closed to it more than it introduces competition to the business. The FCC will probably be embodied to fulfill the requirements of Judge Greene's order. Part of this responsibility is the enhancement of protection, competition, and innovation for those seeking to enter this market. One cannot call an environment intensely competitive if it requires a company to have a billion dollars or a thousand lawyers to enter it.

The new system must be a combination of statutory self-regulation and innovation. It must avoid permanently fixed rules, yet it must have rules. Past evidence shows that the FCC issued policy, whereas AT&T set standards. The new standards must come from agencies and groups interested in competition and innovation. The United States requires a sensible and flexible set of standards that will allow an open network and increase competition.

If the FCC or another overseeing agency does not establish standards and find a way to firmly establish competition, then another solution must be found. One possible solution is to have the business community set the standards, because they are the initial users of this technology and are therefore in the best position to define requirements and ensure that manufacturers adopt them.

The government's role would then be to establish policy, to extend these requirements for the advancement of telecommunications in developing countries, and to work for international standards. International committees are working hard on these standards, but the technology is moving faster than the standards. If the standards could be derived from the requirement specifications of private networks, this process could be improved. Countries with the strongest economies should experiment with new techniques so that developing countries could invest in well-tested technologies. Countries with strong economies (United States, Japan, Germany) have manufacturers vying for this private network and government business in order to establish themselves as world leaders and to be near the customers with the greatest and most farsighted demands.

Another aspect of this growth will be determined by how local and state governments and the federal government approach telecommunications. For example, Houston is representative of many city governments. Houston must improve its transportation system, or no one will locate in the congested downtown area. The city proposed to spend $100 billion on a rapid transit system that no one wants. It would be better to spend that money on the world's greatest telecommunication network to allow a business to locate downtown while most of its employees could locate anywhere in the Houston area. The employee would travel downtown occasionally: thus, congestion and slums, normally associated with rapid transit, would be avoided and Houston would be established as a leader in the Information Society. Cities considering similar systems would do well to look at what a super telecommunication network could do. Chicago became a world city due to a transportation network called the railroad, and the world is waiting for a new transportation leader.

The potential is enormous—we are talking about an entirely new society where the transportation system is the telecommunication network, where people are less dependent on cars, trains, planes, or other items of the industrial revolution. They will be able to access any information they want from any place at any time. The opportunities are also enormous: it is not farfetched to assume that 5–10% of a company's expense will be on telecommunications and that this business will replace the automotive industry as the world's largest industry.

Government and business, however, must recognize telecommunications as the wave of the future and promote it at every opportunity. Businesses, universities, and technical societies must cooperate to establish industry standards while governments regulate and expand these standards at the international level.

Where, then, shall we begin the study of telecommunication? It will be more fruitful in the long run if we begin with a study of networks. Networks are the symbol of power within telecommunications, and they represent the function of telecommunication—the transport of information and the connection between two users.

The main purpose of this book is to define and explain the various aspects of the current network that a "telecommunication" person should understand. The most important activity of this person is to plan new services and to understand various products for this network. By planning, I mean understanding in some detail the job to be done, what is feasible for the job, and when to do it.

The function this person performs is called "system engineering" and this person is commonly known as a telecommunication person or engineer. The current lexicon refers to the individual as a communication or telecommunication manager and a host of other titles. For this writing, the term "network analyst" is sufficient.

Expertise in telecommunication is a recognizable quality, although, paradoxically, it has passed unnoticed until the industry was deregulated. In general, it required an expert to recognize expert technique, since the average user knew little of the workings of the network. Today there are many network analysts, mostly self-ordained, offering numerous solutions to real and imaginary problems. Of course, the true experts are all different, some are strong in transmission techniques, some in traffic, some in switching. Yet in many ways the experts are all the same. Regardless of their individual expertise, they all relate to the many factors that make up a network. It is these thought processes that this book attempts to capture—an ambitious, perhaps, impossible task.

In addition to a firm understanding of the technical aspects of tele-communication, the network analyst's job is to identify the environmental characteristics, both physical and operational, in which the system must operate. From this environmental definition and the current communication know-how, the network analyst must formulate a plan that clearly states the objectives and goals for the network under consideration.

Understanding the relationship of the network to the environment is paramount in formulating the plan. For example, a digital network transmission plan is built on a fixed loss objective for transmission. However, if the environment contains some analog microwave that the customer can't replace, then a VNL (via net loss) plan, common for analog systems, must be utilized until the network is totally digital. The network analyst should not be concerned with the details of the transmission plan or the design details of other plans. The analyst's concern is whether the plan is fixed or VNL and whether the end-to-end objectives of the network are being met.

For more complex networks, a simulation tool is extremely useful because it gives insight into the total network operation, allowing the network to be viewed through transmission, traffic, numbering, routing, signaling, and administration and maintenance. These views are necessary before the network is finalized. The simulation should provide some economic tradeoffs before the final plan is determined. It should not select the final plan, but it should support the solution arrived at through iterations without the computer. The network analyst is responsible for selecting the final plan. In other words, the final selection should be based on sound engineering judgment, not a computer printout. The final report on this selection contains a clear statement of the services and economic advantages to be accomplished by the network.

The issuance and approval of the report should give sufficient information for specifications to be submitted to vendors in order to obtain price quotes for the total network or portions of the network, depending on how contracts are let.

The network evaluation must continue through the procurement stage, because a vendor's proposal will most likely change some aspect of the network, and decisions must be made on whether the network still meets the objectives at the projected cost.

Throughout the procurement and installation stages, technical and economic checks and balances must be carried out. The network analyst must see that the system meets the criteria set forth during the planning and network-selection stages.

Some companies rely on vendors to provide system engineering support; at best, this produces a mediocre system that helps vendor sales but does not meet customer objectives. More and more people are becoming aware that vendors, both switching and data processing, are interpreting and arranging the network (to suit them, if I may be so bold!) rather than clarifying it. Clarification must come from an understanding of the users' needs, a determination of the switching configuration for optimum routing, and an awareness of which features and services will benefit the end-user, the customer.

History
of the Information Age

1.1 Telecommunication History

Early Telecommunication

The first switching system was the operator as she provided manual operation for her customers. Switching systems of the past 80 years have been trying to imitate her abilities. Technical literature on switching systems is plentiful but mainly devoted to, and written by, engineers. I will spend time explaining not how a system works but why a system is important. For our purpose, a switching system performs the same functions an operator performed when telephony started—that is, stages within these systems duplicate the operator's functions.

The operator had a selection stage where she inserted a plug in response to a customer request. During this selection stage, she determined the destination of the call or the terminating number. She then went to a connecting stage where she used another plug to connect to the terminating party. Figure 1.1 is a generic block diagram of a switching system performing these functions.

To simplify a complex process, switching systems engineers have designed machines that either combined the selection and connecting stages, or separated them. The original automatic switching system, called step-by-step, used mechanical switches to advance the customer through the selection stage(s) as the digits were dialed. The customer was processed through the connecting stage by a mechanical switch called a connector. From 1910 until World War II (WWII), this type of switching was more than adequate because the basic requirement of a switching machine was simply to switch calls within the local community. Frills started to become important as we entered the 1960s.

**Fig. 1.1.
Principles of a
Telecommunica-
tion System**

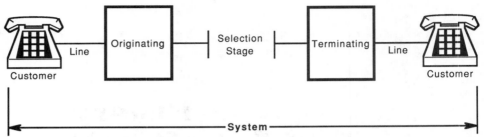

There were many attempts to improve on this step-by-step arrangement, but none were successful until a crossbar switch was used with other requirements. The crossbar switch could not be driven by the customer's digits; therefore, a method was needed to store the digits before making the connection. Also, units were needed to make the connection with the crossbar switch. Early versions attempted to imitate the step-by-step method of connection by going stage by stage. Due to high cost, these methods were not successful. A switching machine (Figure 1.2) called the #5 Crossbar would change the approach to switching systems.

**Fig. 1.2. A
Crossbar System**

First, the #5 Crossbar combined the selection and connection stages to gain some economy; second, it added the ability to bill a call automatically via a switching machine. This billing ability was the first frill for a switching machine and came just after WWII when people became interested in long-distance calling.

The transmission components of telephony did not have the ability to communicate over great distances in the pre-electronic era. An electronic tube applied as a repeater eventually permitted transcontinental as well as international calls. There was, however, little demand for long-distance calling until WWII.

Post-World War II History

The post-WWII boom saw the country demanding not only phone service but good long-distance phone service. They had experimented with long-distance service during the war to someone they loved and were pleased with the results. They didn't understand how the operator was able to get a connection to their son 2000 miles away, but they were happy it could be done. The unremitting movement during this period of the country from a rural life to an urbanized and suburban living style continued at an accelerated pace and brought the phone into every home. The phone was no longer perceived as a luxury but as a tool for doing business and a necessity for maintaining family communication.

The days of the step-by-step switch with its inability to see a connection across an office, much less across the country, were coming to an end. Switches with more flexibility were needed to handle these new horizons. A process called translation was being added to common control systems to provide some of this flexibility. Translation allows a call to "see" the entire office and the paths available before routing the call through the network. The translation would be contained in a device called a processor, which started as a hard-wired device and evolved to a software-controlled computer.

Translation also provided the flexibility to separate the customer's telephone (directory) number from the system (terminal) number and translate from one to the other. Customers could then move across town and have only their system number changed (where the cable pair is located on the system) without having their directory number changed. Operating companies found this feature very attractive in dealing with what was becoming a nomad society. Translation also provided a better distribution of assignment between the heavy traffic users and the light traffic users, a valuable feature in this new era of space-division networks.

The introduction of common control with its translation capability started the idea of a network separated from the controls. In step-by-step, the control and the network were combined into one switch, whereas common control systems separated these two functions but placed the responsibility for the network with the common control. This made the Bell System happy, because it had never approved of customer-controlled networks, which step-by-step provided. Today, customers are gaining control of the network, and it will be interesting to see Bell's reaction.

The network in common control systems was a passive but extremely expensive portion of the system, and a concentrated effort to reduce the cost was made. A. G. Jacobaeus of Ericsson Company in 1950 and Charles Clos of Bell Labs in 1953 took the works of A. K. Erlang (an early telephony traffic pioneer) and others and developed probability of blocking, or nonblocking in Clos's case, for multistage networks. Multistage networks allow concentration or expansion where needed and where blocking probability is permitted. The processors mapped the network and connected the call via an idle path through all the stages of the network, but probability theory provided information on whether an idle path existed frequently enough to meet the system criteria. The network now became an identifiable and costly item.

Crowds of people using the formulas of Jacobaeus and Clos tried to reduce the cost of these networks with innovative and complex methods. Some of the network structures with their many stages and switches and wires would qualify as works of art and were treated as such by many of their inventors. The basic switching device for these networks evolved from the crossbar-controlled switch to a freestanding electromechanical device (reed) controlled by a marker. The marker served as the arms making the connection under control of the brains (processor). The processor was starting to move from an electromechanical unit to an electronic unit as components became available. It was assumed that the network would evolve in the same manner as the processor. But, alas, the electronic device to replace the electromechnical network device never materialized, and the era of the space-division switch was coming to an end.

This change struck network designers hard, for they envisioned their structures as masterpieces. This, together with their esoteric understanding of the accompanying formulas, allowed them to show off or ad lib to management, customers, or fellow engineers at a moment's notice. They greatly resisted giving up this hallowed position. The problem was that the electronic devices could not be inexpensive and still pass a 20-hertz tone at 90 volts, the requirement to ring a phone at the customer premise. Thus, the emphasis shifted to time-division networks and brought an end to space-division networks and the network designer.

Today, switching systems are digital, with time-division networks capable of many more services than those possible with former configurations. In fact, telecommunication switching systems with their computer control and their networking capabilities appear to epitomize the information age more than any other structure. But this is not accidental, because the Bell System contributed greatly to the start of this age.

1.2 Start of the Information Age

Background

An important turning point occurred in industrial civilization during the years immediately following WWII. This turning point, which would affect the entire world, consisted of three independent events, all occurring in the northeast United States, the intellectual center of the country at that time. These events would merge some 20 years later into what is now known as the Information Age or the Third Wave. It is said that humankind has been through three eras, although most people (including authors on the subject) would be hard-pressed to relate when each of these eras began. The Agricultural Era probably started when people decided it was necessary to care for their offspring for more than one season and established a home. When this happened may never be known, but it did illustrate the focal point of the era—the family. If anything epitomizes the Agricultural Revolution, it is the emphasis on the family and with it the relationship of people with the land.

Although more current, most people would not be able to answer the question on when the Industrial Revolution began. One theory is that a small battle in India started the Industrial Revolution. This battle was not a triumphant coup or an ingenious offense destined for the pages of history. It was a minor clash between the forces of England and a ragtag group of Indians from the city of Plassey. By his victory at Plassey in 1757, Robert Clive gained Bengal, the richest province of India, for England. This victory provided sufficient resources (diamonds) to finance the expansion of Indian cotton goods by the development of tools to improve production of these goods. This epoch-making event has gone virtually unnoticed by the ages, but, even then, it illustrated the main ingredients of the industrial era—labor and capital.

The Industrial Revolution shifted the emphasis in this era from the family to employment and the learning of skills. The latter led to a greater interest in education. In addition, the social trajectory of the industrial era would lead from manufacturing and trade to the respectability of education and business.

The Industrial Revolution aided production and, in most cases, lowered prices. This era peaked in the 1960s when the baby boom of the 1940s developed negative feelings toward money and social status. But the seeds for a new era would be planted prior to this time, in part stimulated by the battles of WWII and an attitude following that war.

It would not be an exaggeration to say that after WWII, technological progress far outweighed all that had preceded it since the dawn of civilization.

Before WWII, the function of the scientist had always been to explain the laws of nature, whereas the function of the technologist was to bend these laws to the will of man. Within this process, the Industrial Revolution had indirectly sponsored the Scientific Revolution by supporting education as a means for training individuals with skills needed in the business community. Almost without exception, the scientists belonged to the universities (or were independently wealthy). Usually the scientist was encouraged to work on experiments that would help grow crops or mend bones and to convey this information to society. A prime example of this was the invention of the x-ray by Roentgen. Heralded immediately by the press, it was, within weeks, being utilized by everyone in the scientific community to examine broken bones without surgery.

Electronics in this era had started with the research of Hertz, Thomson, Edison, De Forest, and Marconi. Hertz conducted experiments on electromagnetic waves in 1887. He also observed the photoelectric effect. In 1897, Sir Joseph John Thomson, in England, demonstrated the characteristics of the electron.

Edison had noticed that when a metallic plate was sealed into the glass bulb of his carbon filament lamp a mysterious current flowed between the plate and the filament. This phenomenon is known as the "Edison effect"—a stream of electrons destined to revolutionize electrical inventions, including radiotelephony, wireless telephony, broadcasting, radio, and television.

In 1914, Lee De Forest's three-pronged electrode tube was used as an amplifier on the wire circuits connecting New York and San Francisco, making transcontinental telephony possible and ushering in the age of the electronic tube. Marconi was able to transmit information with radio waves over great distances.

Edison also was responsible for bringing scientists and technologists together, often bragging he could buy any mathematician or scientist. This relationship was embarrassing to many scientists, but it was the beginning of industrial research. A better working relationship would be formed at Bell Labs when it was founded in 1925 to do industry research for telecommunications.

Bell Labs was started with a charter to do research on virtually anything, because the only stipulation was that the subject be related to telecommunications. Within the Labs in the 1930s, a philosophy of change existed with regard to electromechanical switches, which controlled telephone exchanges. Marvin Kelly, later to head the Labs, saw that the electromechanical switches were too large and inflexible for the demands of communications. He asked William Shockley to investigate the new electronics to determine whether these devices could be used in switching. Shockley's initial attempts failed, and additional efforts were interrupted by the war. At the same time, Alec Reeves, an English scientist working in France, had patented a pulse code modulation (PCM) scheme that used binary logic for transmitting telephony signals. His work had no practical application at the time and was soon forgotten with the war. The depression of the 1930s had a debilitating effect on new ideas, since the public had little interest in changes.

The start of WWII changed that attitude and united the disciplines of the scientist and the technologist.

World War II was the first technological war. The scientist and the technologist were to work together on some of the greatest projects ever undertaken, culminating in the Manhattan Project. Included were many electronics applications from radar to the first microwave system, so General Montgomery could communicate with the front without being there. The leadership in these electronic developments was shared by England, Germany, and the United States, but only one of these countries would emerge from the devastation of WWII with the country intact and an optimistic view of the future in electronics.

England, after the war, took a pessimistic view of the future, as illustrated in the works of Aldous Huxley and George Orwell. Huxley wrote his *Apes and Essence* in 1949—one of the first books on what the world would be like after WWIII. The setting of the novel is 2018 (several years after WWIII), and the world is ruled by apes, a familar theme for later books and movies on the same subject. George Orwell wrote his famous *1984* in 1948, published in 1949, with a bleak outlook for humankind—a totalitarian society ruled over by a Communist-like party that used communication to dominate people. Orwell, for his tremendous insight into the future of communication and television, saw these as evils that would be used against man and not for him. But in America the mood was that anything could be conquered because everyone was working together. America had emerged from the war with the country intact and optimistic. Science also was taking a leadership role because many of the schools had sufficient talent to foster a homegrown community of re-searchers. The center of this activity was Harvard, MIT, and Princeton, although the University of Chicago was providing extraordinary results.

Many of the ideas that were pure theory before the war were now reality. Their work in the war gave scientists an exalted position, removed the pure academic cloak from their shoulders, and gave them access to money. Not only was the government interested in pursuing applications of science, but industry also was looking for change.

The stage was set for a new level of attainment.

In 1945, Arthur Clarke wrote his famous article on satellites. Many would include this among the new Information Age discoveries, but satellite communication is really an extension of the terrestrial micro-wave communication and, although unique, is not revolutionary.

But in the Northeast, a group of scientists and technologists would forge and impose a technology that would have not only a short-term influence on the world but would in the long term change forever the dependency of people on the products of the industrial era. It would not destroy the industrial era, but would optimize its use with ideals and make a smaller and better world that could communicate and share information much easier and faster. The world's reach and grasp would now merge.

The Information Age grew out of the marriage of the computer and telecommunications to form the basis of modern society. The telephone network was well established and one of the greatest machines available. If the power of the computer could be added to this network, people could have access to any information they needed. One of the first to understand the potential power of the computer was John von Neumann, who had helped to solve some of the most difficult problems of WWII.

Architecture

One of the most brilliant men of the twentieth century was John von Neumann. At 24, he invented "game theory" from observing the players in a poker game. Game theory is a mathematical method of determining strategies in areas of uncertainties. It is used extensively in economics and is basic for much of the Cold War strategy.

Von Neumann was also a legendary figure in any circle in which he associated. He was a very colorful, absent-minded professor, who once ran into a tree and accused the tree of stepping in front of him. Von Neumann came to the United States from Hungary in 1930 as a guest lecturer for Princeton. In 1931, he joined the faculty of Princeton and made the United States his permanent home. Before and during the war, he became interested in the applied fields of mathematics and physics. His applied specialty was intricate hydrodynamic problems, especially the interaction of shock waves.

He spent many years before WWII teaching and doing research at Princeton on mainly theoretical problems. As the war approached, he became involved in applied research, particularly in hydrodynamic problems requiring complex solutions. World War II saw many new inventions of warfare, not the least of which was the atomic bomb. Von Neumann was one of the scientists who would occasionally disappear into the West during the Manhattan Project. He was also interested in radar, long-range rockets, infrared detection devices, and complex mathematical problems of military operation and code breaking.

Toward the end of the war, the Moore School of Electrical Engineering at the University of Pennsylvania was developing a computer to assist in some of the complex calculations required for the Ballistic Research Laboratories of the Army. The concept of a large computing machine or number cruncher was originally conceived by Charles Babbage in 1833, but serious work did not begin until the problems of WWII became so complex that standard methods were too long and laborious to handle them.

Howard Aiken built a prototype of an electronic calculator using electromagnetic relays, and later worked with IBM to build the first commercial machine for that company.

These early versions were monsters that would only perform one function at a time; that is, the input, the calculations, and the output were performed serially.

At that time, an acquaintance of von Neumann encountered him at a train station and invited him to observe the machine at the Moore school.

Von Neumann caught the significance of this development right away, and he threw himself into the study of it. He appreciated the fact that the machine was trying to take input, do calculations, and present output within one control mechanism. He decided two things about the computer: (1) Digital computers would prevail over analog computers, although the subject was hotly debated. (2) The machine must be capable of processing information. The latter was of greater importance because it meant the computer must have a brain and a memory. The memory would contain data to be processed, and the brain would have instructions that would make modifications possible and separate the input from the output. In other words, the three functions of the computer could operate separately, resulting in a faster, more efficient, maintainable machine.

Von Neumann wanted the computer to operate like the human brain, so he studied neurology and asked neurologists and psychiatrists to help in the simulation. Von Neumann had been influenced in this structural design by Alan Turing, a young Englishman interested in the development of a universal machine. The brilliant and enigmatic Turing adds the element of intrigue to the discovery of computers. Closely guarded by Churchill during the war for his code-breaking abilities, his work is still classified. After the war, he became depressed and non-productive and died mysteriously in 1954 from poison.

The idea of a machine performing all the laborious calculations for the mathematician made sense to von Neumann. His involvement in the Manhattan Project made him recognize the need for the United States to maintain the lead in the arms race over the USSR, and the machine would assist in this goal. He and a select group of engineers built an experimental model after the war for the Institute for Advanced Study. Although that machine had numerous problems and was never success-ful, von Neumann tirelessly preached his cause to the government for funds. These lectures and presentations would form the guide for the computers of the 1950s and 1960s. This same architecture is employed today, and we can thank von Neumann for moving the machine from a number cruncher to a data processing unit.

Von Neumann never received the recognition he deserved, because the computer built at the Institute never achieved any success. He also suffered from "McCarthyism" when he vigorously defended Robert Op-penheimer at a Senate hearing. However, von Neumann architecture is still with us, although you do occasionally see an advertisement for a non–von Neumann computer, which, in itself, is a compliment to this brilliant man.

Building Block

Many problems of early computers were the components used. The thousands of vacuum tubes were not only highly unreliable, but they generated staggering amounts of heat to be dealt with. The answer to the component problem was being addressed at Bell Labs at the same time von Neumann was attempting to build his electronic computer.

The term "transistor" was coined by John Pierce, a Bell Labs scientist who later would make Arthur Clarke's dream of a communication satel-lite a reality by leading the team that launched *Telstar*. The word was intended to denote the transistor as the dual of a vacuum tube with transconductance. Resistance is the dual of conductance, and transresistance would be the dual of transconductance—hence the name transistor. Figure 1.3 shows the three men who received the Nobel Prize for inventing the transistor.

Fig. 1.3. The Inventors of the Transistor—William Shockley, John M. Bardeen, and Walter H. Brattain

After the war, Marvin Kelly (Fig. 1.4), who had directed the study of semiconductors, had become executive vice-president of the Labs and made extensive organizational changes to emphasize solid-state physics as a way to improve the semiconductor progress. As part of these changes, he made William Shockley director of semiconductor research. Shockley built a group that returned to the same type of experiment he was conducting before the war.

One of the first physicists Shockley hired was John Bardeen, who provided rigorous proofs for the physical phenomena Shockley was constructing. Bardeen would eventually be awarded two Nobel prizes in physics: one for the transistor (with Shockley and Brattain) and one for his work on superconductivity. When Bardeen joined Bell Labs, he worked with Walter Brattain, who had been conducting experiments on semiconductors for almost 20 years.

Bardeen could not only explain the experiments being conducted, but he also developed his own theories on semiconductors. One theory discussed the phenomena of the surface of the semiconductor-trapped electrons so that their flow could not be modulated. This differed from Shockley's approach and led to the initial transistor.

Fig. 1.4.
Marvin J. Kelly

Aided by Shockley's work before the war, and with Shockley as director, Bardeen and Brittain conducted experiments that confirmed Bardeen's theory. Later, the device conceived by Shockley would also be confirmed and would prove easier to manufacture.

These men also understood the physics of the transistor. They built, within several days of the invention, an amplifier to demonstrate to the world the birth of the age of the transistor.

The transistor belongs to the world of semiconductors, a science begun with the discovery of quantum mechanics in the late 1920s. The transistor operates like the vacuum tube, except it controls electronics in a solid crystal instead of in a vacuum. Both devices allow the incoming signal to be magnified several times and passed to a collector. However, the size, power requirements, cost, and life expectancy of the transistor make it far superior to the vacuum tube.

The transistor allows matter to amplify electric current without a vacuum, thereby offering tremendous advantages over the vacuum tube. The 1948 *Scientific American* article on the introduction of the transistor contained quite a prognostication when it stated that the transistor "can be made almost vanishing small . . . about the size of an eraser on the end of a pencil." Today, tens of thousands can fit on a chip the size of an eraser, and, for all practical purposes, the transistor has vanished from the naked eye. It is, however, the prime building block of this new age and possibly the invention of the twentieth century.

Besides offering an alternative for the vacuum tube, the transistor began the era of semiconductors. Microchips, memory chips, gate arrays, and so on, all evolved from the concept of the transistor. But to help these devices make their way into the electronic world, a mathematical concept for transmitting information was needed. The public announcement on the transistor was recorded in July 1948. That same month an article by a Bell Labs mathematician, Claude Shannon, on information theory appeared in the *Bell Systems Technical Journal*. This article provided the theory necessary for the new science.

Theory

A structure was needed for this new age, and, as usual, mathematics would be employed to provide control and interfacing rules for the various groups. Mathematics was to be the foundation for this new science. Shannon's article "A Mathematical Theory of Communication," included a theory for transmitting information and for calculating allowable bandwidth, and laid the foundation for modern information theory.

The French mathematician Jean Baptiste Fourier discovered that any waveform can be represented by a set of simple sine waves spanning a band of frequencies. Thus, bandwidth, the range of frequencies of signal waveforms, was discovered and is the present mode of communication. In the human ear, the cochlea sorts out sound waves into sinusoidal components of different frequencies. In sight, the eye interprets light waves of different frequencies as different colors. Both operations are done by filtering—that is, the rejection of frequencies other than the ones to which a response can be made.

The frequency spectrum is the cornerstone of communication, since voice, radio, and television are all transmitted and received via various ranges of this spectrum. To develop a theory of information for this vast array of frequencies, we must avoid the meaning or content of the message and only deal with the amount of information the message conveys. In addition, the theory can provide answers on what facility (bandwidth) is required to process the message. This is what Shannon did. He went beyond the message and looked for the smallest unit of measure for information. This unit of information has come to be called the *bit*.

In the transmission of information and in the integrated-circuit (IC) business, the bit is the main element of information. A bit normally has two states: a "1" to represent the presence of information, and a "0" to represent the absence of information. When engineers speak of LSIs with 10,000 gates, they mean 10,000 devices that can have a state of 1 or 0. When they speak of digitized voice, they mean the ability to sample the voice 8000 times a second and convert the information to 1's and 0's for transmission to the destination.

The sample rate of 8000 times a second for human voice is derived from the fact that the human voice, from Fourier's time, has a range of 4000 Hz (cycles per second). To derive an intelligent representation of voice, we need a scan rate of twice the range. During each scan, the voice or message sine wave is observed, and the position on the sine wave is noted. This position is then converted to an 8-bit code for transmission in the network.

Claude Shannon's work (Fig. 1.5) developed a theory for this transmittal of voice or other information and is now recognized as the definitive work on the subject. His two-state position device was the relay in those days, but gates can easily be substituted for relays without affecting the theory. He understood that the channel for transmitting this information was merely a medium or conduit, and that any message can be broken down to bits and transmitted on a conduit if there is sufficient bandwidth. Figure 1.6 is a sketch of a general communication system proposed by Shannon.

An information source produces a message intended to be communicated to the receiving end. The transmitter is responsible for presenting the message in the proper format to the channel. In digital transmission, this would be 1's and 0's. The message is transmitted via a channel to the receiver, which converts the message to a format understood by the destination. The noise sources are items that interfere with the transmission of the message.

This theory stated that the analog (original) format of the spoken message was not necessary for understanding the information. This led to a reformatting (digital) of the information and eventually to a new telephony, which included a better understanding of the relationship between bandwidth and transmitted information. Before the war, telephony and radio had moved apart, but the necessity of communicating with moving battlefields brought these two sciences together. Shannon's work

Fig. 1.5. Claude E. Shannon

Fig. 1.6. General Information Channel

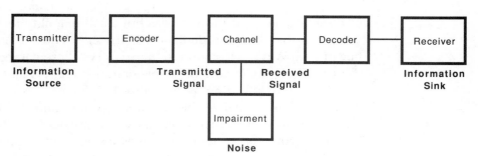

united them with other methods of transmitting information by showing the relationship of information and bandwidth. This relationship was originally noted by Nyquist, a Bell Labs scientist, but it was Shannon who made it part of information theory.

Although Shannon made a powerful contribution to our understanding and practice of communication, he has not received the accolades he is due. Even in the recent, well-written book on Bell Labs entitled *Three Degrees above Zero*, the principle reference to Shannon was his ability to ride the unicycle and juggle balls at the same time.

Present and Future

The Information Age now had the components, the architecture, and the theory necessary to begin a new epoch for humankind, but it would take several years for these elements to merge. The transistor never received its proper credit because it was first associated with the transistor radio and the spread of rock and roll music. The transistor did receive credit from the computer industry because they saw a way to replace the vacuum tube. However, the best commendation came from the manufacturers of the building blocks of electronics, since the transistor was the forerunner of the IC, LSI, VLSI, and the next generation of components.

The appeal of the Information Age is the end of the strong dependency on the items of the industrial age. Improved communication will be the result of this era as language barriers start to disappear. Since the 1960s, there has been a turning inward for personal satisfaction and a turning away from the need to have the latest gadget or the best car. This inward turning depends on the assumption that information, and with it knowledge, will be available for the individual when he or she needs it. This implies an eventual network interconnecting computers and data files as well as voice to allow access to any information in the world when needed.

For America, the Information Age represents not only the discovery of the ingredients of this new era but also the ability to implement them and meet the changing demands of the age. The scientific growth that began with WWII will continue to expand as these components grow and the requirements for scientific leadership is still evident.

We are only beginning to realize the impact this new age will have. It eventually will change the education system and the way we work. It will allow the individual a significant increase in self-realization and increase economic productivity in work and education. Examples are already available as the electronic spreadsheet replaces the number crunching of the accountant. We are capable of overcoming our fears of the electronic future as more programs become available that improve productivity and our understanding of these programs increases.

The Information Age already is inundating us with uncontrollable deluges and eruptions. It is apparent that the mechanisms of learning and working will sustain a tremendous upheaval, hopefully for our betterment. It appears proper to honor the founders of this change and to recognize them in the same manner as the French honor the Curies or the English honor Newton—as true pioneers and people who have changed our world.

1.3 Basic Elements of the Network

Engineers and entrepreneurs are being attracted to the telecommunication industry and away from the computer business. Why? What do these people see in what others are viewing as pure chaos? Of course, they probably attended some cocktail parties where a doctor or lawyer cornered them and carefully explained how they gained access to information halfway around the world in seconds. Or they had a car dealer explain how they hook a telecommunication gadget on their computer to allow them to order a car directly from the factory.

This sudden praise of telecommunication has caught the attention of a number of people. For years most people picked up the phone, dialed a number, and had a conversation without thinking of how the phone company did it. These people are now confused by the changes going on and their larger phone bills. Others are attempting to understand the network and how it can be manipulated for features and services.

What then is a telecommunication network and why is it so important? The latter will take some time. The former will be dealt with here.

First, a telecommunication network is composed of three major elements: switching, transmission, and customer-premise equipment. A network is a system composed of these elements, although there are other elements that could be considered (such as power and signaling) to totally understand the system.

The basic arrangement of these three elements is shown in Figure 1.7. The customer represents the customer-premise equipment, line and link compose the transmission equipment, and switch 1 and switch N represent the nodes closest to the parties engaged in a conversation. The originating party initiates a call by lifting the handset, which alerts switch 1 via the line. Digits relating to the destination are transmitted to switch 1, which interprets the code and begins routing. This routing may comprise a path in switch 1 or several switches, depending on the destination. Once the call reaches switch N, the called party is alerted, and after answering, a conversation is begun. The final connection is via the various line and links throughout the network. This connection can be

Fig. 1.7.
Arrangement of
Communication
Elements

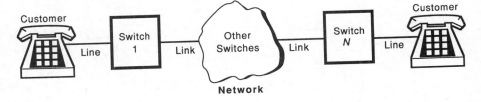

copper wire, microwave, fiber optics, satellites, or any media suitable for high-quality transmission.

All the elements in the network are evolving at a rapid rate due to our new technologies. The customer-premise equipment is moving from a simple phone with copper wire to the switch to a combination phone/computer with fiber optics to the switch. The whole transmission media is starting to be dominated by satellites and fiber optics, both of which were nonexistent 20 years ago. The switching systems are just beginning to introduce the first generation of digital switches into the network, with promises of many generations to come.

The basic elements of a call, however, remain the same, with codes to identify the destination while transmission and switching operate together to gain that destination with quality service. This is important because the technology will overwhelm anyone attempting to understand the network. If you always go back to the fundamentals, the various arrangements will be similar.

More details on network elements are presented elsewhere, but it is essential that you remember the interrelationship between these elements. The rest is easy.

2

Switching

Every day we encounter situations involving switching: for example, a traffic light and a railroad crossing. We pay little attention to this science and merely accept it as part of the daily routine. We pick up the phone, dial a number, hear a soft hum and then ringing. During this soft hum, switching systems have taken control of the call and, from billions of combinations, selected the proper party based on your instructions. The best place to start the study of telecommunication networks is with switching.

2.1 Switching Elements

The basic switching systems have remained architecturally the same over the years, but the speed of these machines has changed substantially. Speed is vital to a switching machine because there is a correlation between the speed and the number of calls—and, consequently, lines—the machine can handle. The operators took seconds to handle a call. The step-by-step and crossbar switches took milliseconds (1/1000 of a second) to handle a call. The semidigital (which will be reviewed shortly) machines operated in microseconds (μs) (1/1,000,000 of a second). Today's fully digital machines can operate in nanoseconds (ns) (1/1,000,000,000 of a second). If you are wondering how fast a nanosecond is, it is equal to the amount of time it takes light to travel 1 foot.

The call-processing capacity of switches can be described in terms of speed and number of operations per call. Speed, in addition to handling more calls, provides faster seizure of links between two offices and reduces the probability that both ends are connected simultaneously to the same link. Fewer operations per call, together with speed, allow a machine to have more features and services while still processing the required number of calls. Before we discuss these features and services, some definitions of terms may be helpful in understanding switching systems:

Switching The process of connecting appropriate lines and links to form a desired communication between two station sets or equivalent units.

Switching system The unit for connecting lines to lines, lines to links, links to lines, and links to links.

Switching stage A portion of a switching network to serve lines, links, other stages, or combinations thereof.

Switching network Switching stages and their interconnections within a switching system.

Concentration A switching network (or portion of one) that has more inputs than outputs.

Expansion A switching network (or portion of one) that has more outputs than inputs.

Controls The equipment shared within a switching system to accomplish a function or functions during particular periods of a call.

Figure 2.1 shows the relationship of these elements in switching.

2.2 Switching Systems Operation

A switching system serving a group of customer lines within a particular area is known as a *local central office*. Switching systems also can function as PBXs (private branch exchanges) or tandems. Central offices usually are located near the geographical center of a group of customers so that the lengths of the customers' loops are as small as possible. The size of a customer's cluster for a central office varies from 100 to 10,000. This latter figure is the number of customers that can be served by the four terminal digits of the office code. The seven-digit office code really is a cluster of three digits representing offices and a cluster of four digits denoting terminal numbers. Certain buildings and switches can handle more than 10,000 lines, and they are referred to as *multicen-*

Fig. 2.1. Basic
Elements of
Switching

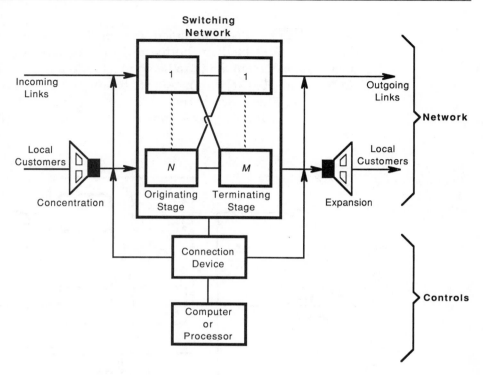

tral offices or exchanges, with each central office having its own three-digit office code.

Calls within a central office are processed by its switching system, whereas calls between offices normally are handled by links (or trunks) if sufficient traffic exists. Obviously, it would be impractical to have links between every central office in a large network, so intermediate offices called *tandem* or *transit offices* are employed for routing under these circumstances. A tandem office serves central offices in the same manner as a central office serves customers. The linking arrangements for a central office consist of direct links to other central offices where they can be economically justified, and links to tandem offices.

Certain offices combine the functions of the central office with the functions of a tandem and are called *toll offices*. This arrangement has been quite effective in rural environments, where a large office serves various small offices and acts as the window to the telecommunication world for the customers in the area. These toll offices are somewhat more complicated to develop, since the billing, switching, and division of revenue (the way the money is divided on toll calls) are different, depending on whether the call originated locally or from one of the rural offices.

The system architecture of a central office is basically the same regardless of whether it functions as a local office, toll office, or tandem office. Figure 2.2 is a simplified view of this architecture. The lines and links are connected on both sides of the network, which allows an incoming call from a link to be connected to a line, and an outgoing call to be connected to a link. All of these operations are controlled by a processor. There are three critical functions of a switching system: terminations for the lines and links; network; and processor or controls.

Fig. 2.2.
Telecommunica-
tion Systems
Configuration

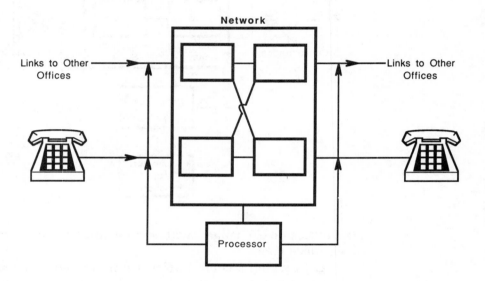

If the call is destined for a customer located on the same switching system, that system will complete the call. If not, the call is routed to the office where the called customer is located, either directly or via tandem offices. A tandem office can be used for long-distance calling or in metropolitan areas where there are many central offices.

Switching calls within an office can be done in various ways. Two of the most popular are step-by-step and common control. A step-by-step switch moves from stage to stage under control of the customer dial without being aware of the total routing within an office. Common control switches store all dialed digits before routing and are aware of the status of the various stages within the network of the central office before routing the call. This arrangement provides more options for the call to reach its destination.

For calls outside its switching system, the common control system routes from office to office without being aware of all options in the network. Common control interoffice signaling (CCIS) gives the call the ability to see the entire network before call routing is started.

Time-Division and Space-Division Switching

Switching stages shown in Figures 2.1 and 2.2 can be either space-division or time-division switches, since the routing strategy is equivalent for either network structure.

Time-division switching had for years played an exceptional role in transmission systems from the time it could achieve 24 conversations over a pair of wires, where only two conversations were possible previously. It also provided a supremely concise elegance to years of work on information theory and the new electronics. Much of what is employed in electronics today goes back to the work of Claude Shannon. Time-division switching had been considered too expensive for switching systems, but the decreasing cost of electronics has changed this position. Time division actually adds another stage to the various stages already in the network and necessitates the use of electronic elements for switching. These two considerations can reduce the number of equipment frames needed for the system's network from 20 to 1 in a large office. How does it handle the ringing of the line with its requirement for 20 hertz (Hz) at 90 volts (V), which was a problem for the original electronics? It bypasses the system's network. A separate connection is made to the line for the ringing.

Blocking through time-division networks is aided by the fact that time is another switching stage. Consequently, the controls find many more options in searching idle combinations. This switching was additionally helped by the work of Hiroshi Inose at the University of Tokyo on time-slot interchange. Time-slot interchange allows one side of the conversation to independently find an idle time slot in a memory unit and store the information in that unit. The receiving half of the conversation would then seize the information during one of its idle time slots. Before this work, both ends of the conversations would have to match idle time slots. This new scheme greatly enhanced the blocking probabilities through the system's network and made time-division switching economically competitive. The two main approaches to time division were pulse amplitude modulation (PAM) and pulse code modulation (PCM), with PCM eventually getting the nod since it is compatible with the binary coding used in computer and data transmission. An explanation of PCM is given in Section 5.7.

Time division has brought together two elements of telecommunication—transmission and switching—that had long been separate and distinct. Switching was only concerned with getting the call from one end of the office or the network to the other, whereas transmission was concerned with the quality of the voice and its improvement over long distances. Time-division PCM presents a common ground for both switching and transmission in switching the call end to end through the network. The advent of time division eventually will reach the customer's terminal or handset, and when this occurs, we will have a fully integrated network. Before this happens, the ringing problem, the two-wire customer loop, and the carbon transmitter will be replaced or some novel solution will be found to integrate them into the four-wire PCM network. The signaling portion of the network will be separated from the transmission and switching by the use of CCIS or equivalent equipment. Eventually, this separation should work its way to the customer's loop.

The ubiquitous two-wire network we know is going to slowly evolve to a four-wire transmission network with a separate signaling network to satisfy future needs. The telephone operating companies, however, argue that they can't replace such a large investment for the special requirements of a few. The other side of that argument is that the special few will find a way to bypass the present network, and the operating companies may find themselves without customers. It will be interesting to see which way the network restructures itself during the next few years.

2.3 Switch Blocking

The switching network is the framework for any switching system, and its configuration can, to a large extent, determine the nature of a system. Modern systems with time-division and space-division networks are not as dependent on the switch configuration as space-only networks. Space-division networks still provide the best method to illustrate a configuration. Assume you would like to interconnect eight customers as in Figure 2.3.

For eight customers there are 28 combinations. The convenient mathematical way of expressing this arrangement is $n(n - 1)/2$, where n is the number of customers. This arrangement requires almost half a million wires for 1000 customers and is virtually unmanageable. For the first 70 years of this century, engineers and mathematicians have been seeking the ultimate solution to the configuration problem, since no single configuration was ever universally accepted. In many switching systems during this time the network accounted for approximately 50% of the cost of the system. We had stage-by-stage networks, folded networks, nonfolded networks, concentration, expansion, distribution, and

Fig. 2.3.
Interconnection
of Eight
Customers

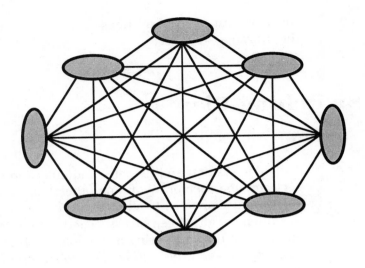

numerous other schemes to reduce the cost and improve the service through the system.

All of these configurations had one thing in common—blocking. That is, a certain probability always existed that the customer you requested was free but could not be reached because the right combination did not exist through the system's network. Some nonblocking networks were created for military applications but were considered too expensive for commercial use. The inventor of the accepted nonblocking approach, Charles Clos, did tell an international group of telephone network experts that if the manufacturers would adopt nonblocking networks, the equipment cost would be slightly higher, but it could be offset by getting rid of the expensive network experts.

The introduction of time-division networks and program- or software-control systems has essentially made the networks nonblocking. In other words, digital switching allows networks to add time as a dimension to the configuration. This greatly reduces the space portion of the network yet increases the combinations for completing the call, thus substantially reducing network equipment, cost, and blocking for the resultant configuration.

In time-division switching systems the processor seeks a connection to any of the space configurations where an idle connection exists to the terminating customer. A typical number for a modern time-division system is 192 different space configurations. The cost of the system's network using time configurations and digital switching has reduced the network portion of the system cost from 30 to 40% to about 10%. With this new cost factor, variations on the network make little difference in the overall system cost.

2.4 Switching Plan

General

Switching systems and the telecommunication network exist to interconnect any line (person) with any other line via a direct connection. The line can be connected to a home, office, pay station, or whatever as long as a unique address can be associated with it. With a unique address for the line the network, using its complex web of switching, transmission, signaling, and controls, can connect to any other line via a dialing pattern without the assistance of an operator.

Rules and restrictions, however, are necessary to allow the present network to manage such a large task. The present network is controlled chiefly by switches (nodes) that are interconnected by links with combined transmission and signaling arrangements to assist this control. The ultimate arrangement will separate these link functions and provide the customer with control of the network. Strategies for these new configurations must be considered in any planning.

Numbering Plan

The numbering plan developed over the years for the network consists of three parts:

1. A three-digit numbering plan area (NPA) code that identifies a geographical area
2. A three-digit office code that identifies a central office within that area
3. A four-digit station number that identifies the location of the line within that central office

This theoretically gives 10 billion combinations to work from, but there are certain restrictions that greatly reduce that number. For example, the NPA code actually looks like this:

$$N1/0X$$

where

    N is any digit from 2 to 9,

    1/0 is the digit zero or 1,

    X is any digit.

These rules, which are necessary to simplify routing, reduce the number of possible NPA codes from 1000 to 160 ($8 \times 2 \times 10$). The numbering plan is also restricted because every state is given at least one area code and certain codes have been reserved for special applications (e.g., the 800 code), thereby reducing the number of codes to 152.

The office code has some of the same restrictions. For example, the office code can't conflict with the NPA codes. (Recently this restriction has been modified for large areas.) To accomplish this, planners let the first two digits be any digit from 2 to 9, and the last digit can be any digit. This allows 640 office codes ($8 \times 8 \times 10$). The combination of area code and office code had a potential of a million codes (six digits), but it has been reduced to 77,280 combinations (640×152). This present arrangement is starting to cause problems in congested areas. The complete introduction of stored program control (SPC) switches will allow some of these restrictions to be dropped, thus making more combinations available.

The terminal digits (the last four digits of the number) have similar restrictions and cannot serve 10,000 customers or stations. For example, pay stations or public telephones (pay station is a misnomer these days because many of them don't require a coin deposit to operate) are served from the 9XXX series to provide a check digit for the operator on collect calls.

In summary, the numbering plan does two things: It uniquely identifies every station in the network and allows the various switching machines to route the call to the destination by breaking the customer number into a definable set of codes upon which the switches can act. International dialing uses the same principles and adds digits in front of the present 10-digit plan.

2.5   Routing Plan

Alternate Routing

A telecommunication network is normally represented by nodes and links. *Nodes* are sources or destinations of traffic (switching systems), and *links* are communication lines between the nodes. Because it is not practical to interconnect the thousands of nodes with each other, a routing strategy is required within the network. A principal criterion for this routing strategy is to route the call as economically as possible while maintaining an overall acceptable grade of service.

The concept of routing developed for North American switching was based on hierarchical configuration, alternate routing, and end-to-end signaling. Hierarchical configurations assign the various switching offices within North America a position in the hierarchy and a direction for switching calls. This arrangement permitted a routing philosophy that did not depend on any "brains" or processor controls in the switching systems. This was convenient for switching systems that could not store and translate digits during the initial implementation of this scheme. Alternate routing was, and is, a method for optimizing the traffic-carrying capability of the network by using the fewest links. One way to understand alternate routing is to picture your trip home from the office when the way you normally take is placed under construction. Consequently, the way you actually go home is an alternate route. Telecommunications can perform alternate routing within the network based on economic values associated with each route.

The introduction of SPC systems has changed the concept of routing calls on the network. Before SPC, the Bell System virtually ruled the routing of every call on the system. One of the biggest changes to the network was the introduction of digital switching; equally important, however, were the changes in routing. The breakup of the Bell System has turned the network over to the computer and software control, so more customer routing strategies are possible.

Hierarchical Switching

The hierarchical structure that has been around for 50 years and is still the backbone of routing calls in North America is built on the structure in Figure 2.4. Two of the offices are customer-serving offices, whereas the other three levels are devoted to switching links (or trunks) between offices. In North America nine chains, all similar, route the various calls. Each chain has a Class 1 office, and they are all interconnected to complete the arrangement. Calls can proceed up the originating chain and down the terminating chain. A call can cross from one chain to another at any place, but once having crossed it cannot return to the originating chain and can only go down toward the destination in the new chain. This was an extremely clever arrangement to prevent a call from cycling in the network even though the offices were not equipped with common control.

Fig. 2.4.
Hierarchical
Routing
Arrangement

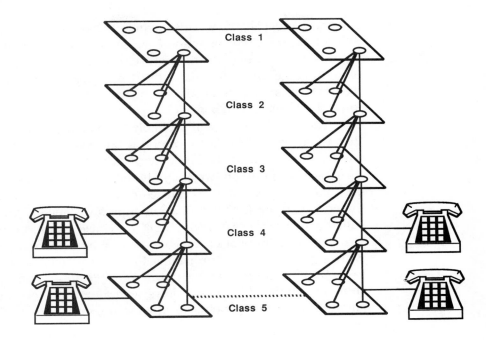

Hierarchical switching with alternate routing offers greater flexibility and lower cost than a nonalternate route configuration. For many years the Bell System administered this configuration and offered compatibility buffers or interface devices for incompatible end offices. This switching arrangement was one of the Bell System's great contributions to the universal service concept.

Note: There are exceptions to these guidelines, but the network has operated this way for years on more than 99% of the calls.

One of the first attempts to employ a more effective routing with common control was the military AUTOVON network, which used a grid pattern to route calls. This pattern did not improve the network efficiency substantially, but it did provide a more survivable network. As a result the public network did not change. Only with the divestiture of the Bell System are we starting to see a change to the hierarchical structure, as more common carriers enter the market. Class 1, Class 2, and Class 3 offices mainly are concerned with switching calls between offices and are now part of the AT&T structure, whereas the Class 4 and Class 5 offices now belong to Bell's regional companies. (I know, this is an oversimplification of a complex problem.) There is still some debate on portions of the Class 4 offices. If we place the Class 1, Class 2, and Class 3 offices in one category and call them "link switchers," we can now have an MCI, a SPRINT, or any other common carrier offering link switching. The routing scheme they use can be hierarchical, grid, or whatever. In fact, this breakup, together with the rapidly decreasing cost of facilities, makes the hierarchical structure obsolete, and we are starting to see an evolution toward a lattice structure.

The link switcher will probably become more functionally oriented: that is, there will be link switchers for operator traffic, for wideband traffic, and for video. The link switcher's network may be reduced from three levels to one or two for national calls. A superswitcher will probably be made available to interconnect all functional switchers on an international level. Three or four superswitchers (I hope they are not called superswitchers) throughout the world would be sufficient.

For local networks a two-level multialternate route configuration with a maximum of four routes for the call will be the dominant approach. This configuration allows very flexible traffic patterns and is shown in Figure 2.5. It also allows several options for interconnecting to common carriers, which will change during the next few years from many to few.

Fig. 2.5. Two-Tier Routing Configuration

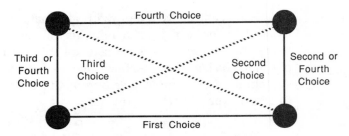

Economical Routing

Changes going on in the hierarchical structure cannot be seen by the customer, but changes in routing are apparent. Customers can now make economics decisions on routing without depending on the telephone company to do it.

I'm sure it's an oversimplification of the routing or switching plan to say that the end office is responsible for routing the last four digits of the number while the other offices are responsible for using the other three or six digits to get the call to that end office. In fact, primary centers, sectional centers, and regional centers should not have any lines connected to them because their function is to route calls through the network without actually terminating any calls. They are also called transit or tandem offices. This does indeed reflect the routing scheme used for the long-distance network, but complications enter the picture when the loss items—call loss and dB loss—are considered. Call loss, or probability of blocking through the network, determines the linking arrangement between the various offices. The dB loss determines the number of links permitted in the connection. I prefer links to trunks because it's a more descriptive word, but you will see trunks or trunking in many articles on telephony.

The alternate routing structure was, for many years, controlled by the phone company and was the only way a customer could route a call. Also, the customer would pay the same price regardless of the route taken or facility used. This changed as WATS, private line, and other toll devices appeared. The first major change that allowed customers to make their own routing decisions was MERS (most economical routing scheme). An example of MERS is shown in Figure 2.6.

Fig. 2.6.
Hierarchical and
MERS Routing

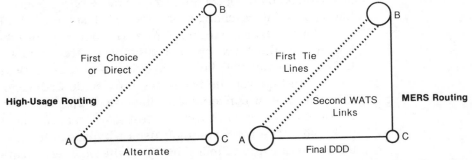

The device is programmed to select the link from the least expensive to the most expensive. Many MERS devices can also provide class restrictions to certain customers over certain links. The MERS option lets the customer take economic advantage of the route selected, which previously was the domain of the telephone company. Deregulation added another dimension to the routing strategy because now the customer could select the carrier for a particular route with overflow to the AT&T network. As toll business competition heats up, even more sophisticated MERS devices will probably be created. Consider the need for a MERS device that will select different carriers for different routes, depending on the time of day. We can expect these carriers to be offering special deals, depending on where they have large number of links. A sophisticated MERS device can be programmed to take advantage of these deals.

One challenge of the MERS approach is to determine the amount of traffic carried on each leg of the routing scheme. This challenge becomes more complicated because not all traffic will be routed through all the legs; some traffic may enter the route selection at midpoint, and some at the last route. The traffic that overflows from previous routes is considered nonrandom, and the traffic first routed on any leg is considered random. Combining random and nonrandom traffic to determine the number of links required is complicated, although programs have been written to help make these calculations.

Other methods are available for taking advantage of the rate structure on special routings. The discrepancy between interstate and intrastate rates has led some companies with networks within a state to drop a line across a state border to take advantage of interstate rates. This technique is known as the "rusty switch." Today these companies should be comparing interLATA and intraLATA rates to see if the same approach will yield savings in this new environment.

Local access and transport area (LATA) is a geographic area within which a telephone company is allowed to handle call delivery. Depending on the state, there can be from 1 to 20 LATAs per state. Calls within LATAs are handled by the local phone company, whereas calls between LATAs are handled by interexchange carriers (IECs), such as AT&T and MCI.

Routing traffic within LATAs also is new because these companies no longer use the hierarchical switching of the former Bell system. The LATAs normally will link to a point of presence (POP) for a common carrier and deliver the necessary information concerning the call.

The use of concentrators has also helped to reduce telecommunication cost, but one of the greatest growth areas has been the concept of bypass. The most common use of bypass is for a business to own a telecommunication facility between the business location and an IEC or a private network, thereby avoiding the handling of most calls by the local telephone company. Bypass has seen tremendous growth since deregulation, and this trend is expected to continue as the cost of facilities continues to decrease. Bypass should be included in any study on routing.

Traffic routing was once the esoteric domain of a few telephony engineers. Now it is common to see new devices introduced to optimize unique routing patterns. Routing telecommunication traffic will be so commonplace that people will be explaining it to their neighbors by the example of the drive home from the office as an analogy.

End-to-End Signaling

For years, the North American telecommunication network has used a connection method known as end to end. With end-to-end connections the total number is passed to each succeeding office, which, in turn, will attempt to switch to an office as close as possible to the destination and pass the number to that office. The call passing through the network may have digits added or deleted from the destination number, depending on the route selected.

The other popular method of making network connections is the originating-register scheme used in most European countries. With this method the originating office maintains control of the call as it is routed through the network. Each intermediate office is given sufficient information to "look" toward the destination and "cut through" to the destination or the next office in the route. If a route to the destination is not available at the intermediate office, the call is returned to the originating office and another route is attempted.

Regardless of the merits of these schemes, they will be replaced by a common-channel signaling method that will allow each switch to look ahead in the network to determine an optimal route for the call and, in many cases, check the status of the line or terminating office.

The current generation of SPC switches, either digital or analog, has link functions converted to software to reduce the complexity and cost of interoffice channels. These functions will disappear as common-channel signaling is introduced. Also, common-channel signaling operates off a separate processor, thereby increasing the real time available in other switch processors. This should result in additional call or feature capabilities for the various switching systems.

2.6 Stored Program Control

The introduction of SPC or computer-controlled systems in the 1960s and 1970s permitted the start of a variety of features on a switching system other than call processing. The #5 Crossbar did have billing associated with it, but, although it was a major breakthrough, it was only the tip of the iceberg as far as features for switching systems were concerned. Stored program control allowed special features—call waiting, call forwarding, abbreviated dialing—to be used. It permitted these features to be implemented with software code instead of hardware, which previous systems had used. Each piece of code would start to define a function that eventually would be carried out in hardware, but the hardware could be used for other functions or services, depending on what the software decided it should do.

It also led to software engineering. The software engineer was the only one who understood the process. The hardware engineer wanted the circuits to perform their functions in an orderly procedure, but the software engineer wanted to keep cycling back to perform the same, or a similar, function using that hardware. Many engineers could not make the transition and ended up designing circuits with more hardware and only token software control. The early software engineer was a true artist in his ability to understand the total picture of the telecommunication system.

The basic structure of the early SPC machines was very simple—computers did everything. They were, however, not called computers because Bell was worried about what the FCC and others would say if they announced that they were selling computers. So they were called processors or SPCs or whatever. The best description for them is real-time computers. Real-time merely refers to the fact that when you want service, the systems must respond to your request within certain criteria. Examples of requests are call for dial tone, digit dialing, and answer. Figure 2.1 illustrates an early SPC office.

With this configuration all calls are controlled by the computer, whereas previous structures would separate particular functions. This led to a new telecommunication skill known as real-time analysis. With the computer controlling all aspects of the calls, a method was needed for determining how many calls a particular computer could handle based on the number of cycles and functions per call together with a call mix. The early SPC machines had problems handling the calls for many of the large offices because the real time per call was underestimated. In other words, the real time per call on paper was estimated at 30 milliseconds (ms) and was actually 100 ms when it got to the field. Even today this is a necessary function with switching machines; in fact, this skill has been picked up by computer manufacturers to determine their machines' limits on terminals and functions.

Real-time criteria is one of the principal items separating one switch from another. The manufacturer should not only be capable of indicating to you the real-time capacity of their switch but also the capacity of that switch within the environment of your network.

2.7 Switch Control

Although the basic control of the switch has moved from hardware to software, the functions performed have remained the same; they are:

Alerting Includes ringing to indicate a call, dial tone to indicate digits can be received, busy tone for customer engaged, and other functions.

Supervising Includes recognition of changes of state, such as a customer initiates a call or terminates a conversation (some of these are occasionally called attending).

Receiver Information from either the customer or another system.

Transmitter Information to a line (busy testing) or to another office (sending).

Recording The information necessary to bill the call or to determine service associated with the call.

Some of these services vary, depending on whether the call is within the public network, within a business, or handled by an operator. With a single computer handling all functions, capacity problems were created from either a timing standpoint or memory standpoint. Without going deep into the subject, a telephone system must respond to your service request within certain predetermined parameters, which qualify it as a real-time system, as opposed to a mainframe computer, which gets to you when it has free time. These real-time parameters and the reliability criteria of a telecommunication system necessitated a unique computer development whenever a new switching system was introduced. These real-time criteria and the need to minimize the investment in the hardware used throughout the call have led to simulating the system software arrangements.

The simulation of the control of switching systems indicated the problems associated with controlling the entire system by one processor. As the system grew and additional services were added, the analysis from the simulation questioned the ability of the processor to process a sufficient number of calls required by large offices.

In addition, the network continued to use electromechanical devices to hold the connection during conversation, although the controls were moving to digital techniques. This was fine for voice, but it did limit the switch as far as data was concerned. Meanwhile, the transmission facilities were starting to use digital concepts between offices, and data networks were starting to emerge. A new generation of switches would be necessary that would employ digital concepts to match the transmission facilities, to handle high-speed data, and to overcome the problems associated with one processor handling all the functions of the switch.

Distributed Processor Control

The availability of the quintessential microchip or computer led the system development away from the total dependency on one centralized processor. Also, the software knowledge was growing rapidly as more and more universities were graduating computer-trained individuals and the press was taking up the call for computer-related careers. The SPC machine was very successful, but, due to the high "getting-started" cost, it was limited to offices with a large number of lines. It also was prevented from operating in very large offices where real-time problems might be encountered.

A switching system cost is divided into getting-started cost and the per-line cost. The getting-started cost covers the computer and a lot of support equipment. The per-line cost covers the individual line plus an incremental link cost. The per-line cost for a computer-controlled system was lower than other systems, but it normally required a substantial number of lines to offset the high getting-started cost. Customers added these to obtain their initial per-line cost; if this cost was high compared to other switching machines, they had to accept the alternative or somehow add many lines to the comparison to favor the computer-controlled system.

To correct these problems, engineers began in the late 1970s to develop the distributed control switch with multiprocessing. This development introduced the concept of a digital system's network for switching calls. This permitted the switching system to match the development going on in the transmission gear as it switched from analog to digital. The approach is shown in Figure 2.7.

The customer line enters the switch domain locally or via one of the remotes. In either case the call is serviced by a peripheral computer that receives instruction from the centralized control for features and services. At the line stage the voice is converted from an analog signal into a digital signal and is maintained in that mode throughout the switch.

This process will be explained later, but for now it is sufficient to say that the digital signal has the same format as data (1's and 0's). The switch network is digital—that is, it uses time as one of its dimensions—which needs much less equipment for any particular size when compared to an analog network. The analog network advantage was the fact that the voice signal remained analog throughout the system. Presently, these two items are offsetting, and there is little cost difference between a digital system and an analog system. The major advantage digital systems have is that most modern developments are in the digital area, and digital systems will be able to take advantage of these improvements. The cost of the device that changes the analog signal of the voice to a digital format has already been reduced by 50% since 1982. Most systems require one of these devices per line, so the per-line cost has been reduced considerably.

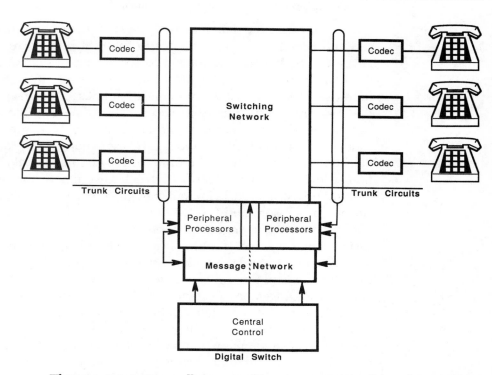

**Fig. 2.7.
Distributed
Processor
Control in a
Digital Switch**

These systems generally use multiprocessor controls, a change from the monolithic processor of the 1960s and early 1970s. The early SPC machines, in addition to the high cost, had problems in incorporating new services and features in the system. Certain manufacturers of switching systems went out of business when the software people told them that the cost of adding certain features to their machines exceeded the original cost to develop the system. The multiprocessor concept appeal is its ability to support small remotes as well as large offices through the use of special processors that are colocated with the customer but derive their maintenance and administrative functions from a central control.

This approach not only improves the cost base for the system, but it also provides a means to pass the new telecommunication features to smaller offices and eventually to all customers.

The multiprocessor control is built to allow different functions to be introduced without disrupting the basic existing network. The new systems are capable of interfacing with digital transmission configurations and processing data, neither of which could be accomplished by the previous machines. Without doubt, telecommunication systems will be required to talk to business systems, data systems, video systems, and other systems in the not-to-distant future. This architecture should allow for interconnections to various configurations without disrupting the existing function.

Also, once the customer has invested in the base unit, the system can add remote units or satellites to the basic structure without incurring an additional getting-started cost. This permits small offices to enjoy the benefits of electronic controls and creates an infrastructure of communication built up from communication links and transmission facilities that permits all customers to receive the same services without being constrained by boundaries. This is really the first important step for establishing a total information network where these large complexes will communicate with one another to create total global telecommunication so that information can flow freely throughout the world.

In this era of distributed control, the relationships between the various processors and the handling of the call are so thick that, often without realizing it, engineers design themselves into a trap that can have serious repercussions for the system. The distributed control is a vast improvement over the monolithic processor, but it is still precarious due to a complex and intricate arrangement between the various processors. Such an arrangement necessitates unique controls over new sophisticated features by the manufacturer, thereby creating a long time lag between the emergence of the requirement to the actual implementation. Is a new generation of switching systems necessary to provide the versatility of the distributed processor with a method whereby customers can add sophisticated features? Such a development would move the telecommunication industry from a service-oriented business to a consumer-oriented business. Perhaps, that is the next plateau for telephony.

Software

With these new systems, software came. The nature of software and software structure had a major effect on system organization, capacity, cost, and ability to incorporate new features.

The uniqueness of the software occurs not in the call-processing part but in the diagnostics section of the product together with the ability to add features. Call processing is the same the world over and actually accounts for only a small portion of the software. It is estimated that over 70% of the software associated with a telephony switch is devoted to diagnostics, which is the area that makes the computer controlling the switch unique from other computers. A typical demand is that the downtime be less than 1 hour in 20 years, or 1 in 175,200 hours. Other computers never approach this level of reliability; to achieve it requires a high degree of sophistication. Programming for these switches will run 200–300 man-years.

To avoid the high programming cost, engineers must design software so that new features and services—many of which are unknown when the original program is developed—can easily be added.

One of the most critical aspects of dealing with vendors is evaluating their software and its capability. High-quality software is vital to meeting systems' high-reliability standards. As software continues to proliferate in the corporate environment, it is becoming increasingly difficult to evaluate its quality. Also, the company owning the software goes to great extremes to protect it from competitors.

Agreements and concessions with the software vendor are highly important to the user, because maintenance personnel will make far-reaching decisions based on their knowledge of system operation. It is not inconceivable that a company's total operation could be shut down by a software error or a maintainer's interpretation of an error. Discussing this sensitive and hard-to-define issue of system fault detection must have the highest priority in dealing with a vendor. This discussion should not center around the features of the system, because here it is unimportant whether the feature is implemented in hardware or software: the concern is the system's ability to detect faults and indicate to maintenance personnel where the trouble is and what the diagnostic action should be.

Software analysis should not be done at the language level, unless some language suddenly demonstrates tremendous ability over the present crop, but on the diagnostic capabilities of the language. Some languages can handle the call-processing requirements and the feature functions of telephony. Most, if not all, programming is structured; no discernible difference will be found in this area. All vendors will tell you their machine can readily add each and every new feature. What they don't tell you is that, in many cases, adding these features involves an infinite amount of money and an infinite number of programmers with an infinite amount of time to create the feature. Adding a major feature to a telecommunication switch can cost more than $10 million. I do not consider that easy and flexible. However, other features may come with very little programming, so it isn't a fair way to judge systems. Judge systems on how well they perform diagnostic checks and maintenance.

2.8 PBXs

For most of telecommunication history, the business community has been served by the PBX. This meant that the connection from the network was through a special number that brought the customer to an operator who routed him or her to the proper party. The advantages of this arrangement were that special features and fewer digits could be provided for dialing within the exchange, and access to and from the business was via an operator who would monitor incoming and outgoing calls.

However, times change, and business started looking for ways to obtain and retain good employees and provide better customers service. One method was access from and access to the network without going through the operator, better known as direct inward dialing (DID) and direct outward dialing (DOD). In addition, for most companies, communication has become a vital part of the business strategy. Organizations must understand communication networks' products and services and how they are impacted by deregulation, competition, or technology.

For business networks, the PBX has long been the cornerstone of the general communications needs, although, depending on the size of the network, centrex and tandem switches also have contributed. Around the world PBXs are being converted to digital technology, because it is rapidly replacing the existing analog version. This digital revolution affects not only the switching systems but also the facilities between systems and eventually the facilities between the system and the customer. The business community must understand this new generation of switches.

Most people associated PBX with voice communication. Its history consists of a private automatic exchange (PAX) that only handled internal calls (a few people would have Bell phones to make outside calls) and the PBX, where the switchboard handled all incoming, outgoing, and internal calls. A private automatic branch exchange (PABX) is a switch that can initiate external calls without an operator and handle internal calls. The more descriptive term for the current generation of switches is PABX, but PBX is commonly accepted. Most PBXs will route incoming calls to an operator.

The term "centrex" was introduced in the 1960s to add confusion to an already esoteric system. Centrex refers to the equipment's ability to operate within the long-distance network: that is, the instrument at the customer's desk has a number that is part of the nationwide system and an internal number. As originally conceived, centrex was divided into two methods—centrex CO, which used equipment from the nearby central office, and centrex CU, which used equipment at the company premise. Centrex provided the features of the PBX (abbreviated dialing and special services) but also allowed direct incoming calls without the assistance of an operator.

The concept of centrex was excellent, although the original implementation of it left something to be desired. Bell decided to charge the centrex line at the same rate a local subscriber would be charged. Suddenly, a business that was paying the local phone company about $1.50 a month per line started paying $6.00 a month for the same line. Remember much of that charge is for the care and feeding of the distribution system to the home and would not be incurred in a large office building. This decision and the ruling that foreign attachments could be made to the network opened the door for Rolm and other vendors to compete in the marketplace. Also, many of the services placed in centrex switches couldn't be offered because of a ruling by the FCC. If the rates are lowered and the current inquiry before the FCC is favorable, centrex will find a market and could be the switch of the future.

The current generation of PBXs is digital switches serving either the PBX or the centrex market with various techniques and sizing approaches. For the most part they are voice-only devices, although some data is handled, normally on a special basis.

The PBXs are now turning to combined data/voice switches in an effort to compete in this emerging market. The drive to interface to ISDN (integrated services digital network) and to combine data and voice on the same switch are the results of a need to share the same transmission facility, to have one instrument at a desk, and to provide the user with many more services. The latter will evolve as bandwidth increases and the user finds that the instrument can be used for more than voice.

Currently, many transmission facilities are employed for data, with very low occupancy; placing both voice and data (and in some cases facsimile) over these facilities will produce enormous savings.

Although the present PBXs are capable of switching data, a new generation of switches is being developed that will have more data flexibility; for example, they may be able to allocate bandwidth dynamically, depending on the requirements of the customer. These capabilities presently are offered only via peripheral devices. The advantages to providing these services within the PBX are that the switch is nearer the customer's demands, and the need to purchase additional equipment is eliminated.

3

Traffic Engineering*

A heuristic model of a stochastic variable is the traffic engineer's way of saying "let's guess at a number." Understanding networks necessitates a study of the amount of telecommunication traffic they can carry in different situations, but this leads to the esoteric language of the teletraffic engineer. We need a practical understanding of this subject to visualize its capabilities. I may know how tall I am and how much I weigh, but until I understand the code of "42 long," I won't know where to look for suits. We need to understand the code of traffic engineering in order to look for switches and configurations in the right place.

3.1 Theory of Traffic Engineering

Introduction

During the early days of the manual switchboard, someone observed that phones were used only occasionally and that providing cords for every possible connection would not be practical. Since those days, teletraffic has become a sophisticated science based on advanced mathematics and complex simulation models.

To determine how well a communication system is operating, we need a relationship between the offered (or input) load to the system and the carried (or output) load. Determining this relationship is the function of teletraffic theory. The formulas used are derived to measure carried traffic at a particular loss probability relative to the offered load. In most cases the formulas also can derive the number of circuits required to provide a particular level of service. The art of teletraffic theory and its

---

*Adapted from *COMMUNICATION AGE* February 1985. Copyright 1985 by Telephony Publishing Corp., Chicago, Ill. All rights reserved.

associated formulas were originally derived from statistics, probability theory, gambling (chance), and have substantially contributed to the congestion theory portion of this science. These formulas are really part of any congestion theory study. Ironically, they were rediscovered during the 1960s by computer engineers working to solve computer queuing or delay problems.

The formulas are basically concerned with one of two criteria: a delay in obtaining service; or a lost call, where the call is denied due to lack of trunks, links, or paths. A formula normally consists of three major elements: the service (or loss) criteria, the amount of real or anticipated traffic on the units under measurement (or offered traffic), and the number of links or service circuits.

Using these formulas, the telephone industry has adhered to high-quality telecommunication traffic standards. These standards define the quality of service the customer will experience when placing a call or gaining access to the network. A traffic engineer's tools range from a method to measure traffic intensity to formulas that would interrelate these items to sophisticated network models. A major concern in traffic measurement is the traffic intensity of the unit under study. *Traffic intensity* is a dimensionless ratio used for determining the congestion or occupancy of a unit. It is measured in erlangs (E), which are named after the pioneer of teletraffic theory, A. K. Erlang. When the unit under study (line or link) is fully occupied for an hour (or the study period), it is said to have a traffic intensity of 1, or 1 erlang of traffic. Thus, a communication line that is occupied for 30 minutes during the hour has an intensity of 0.5 E. A group of 10 links, each occupied for 30 minutes during the busy hour, has an intensity of 5 E (10 links × 0.5 E per link). This measurement can indicate the average number of occupied links, 5 in this case.

The erlang also is useful in deriving the number of calls from a known occupancy. For example, if 20 tellers in a bank have an occupancy of 10 E, each teller is occupied 30 minutes (10 E/20 = 0.5 E) on the average. Therefore, if a teller requires 3 minutes to handle a customer, a teller should handle 10 customers during the study hour.

Centum call second (CCS), or hundred call second, also can be used to determine traffic intensity. There are 36 CCS in an hour, so 36 CCS equals 1 E. The method for determining an intensity measurement using the CCS becomes convoluted and more difficult with the introduction of data. The ideal way to measure telecommunication network traffic is with a good intensity measurement, regardless of whether the traffic is voice, data, facsimile, or anything else. The erlang is easier to use than the CCS, so the erlang lends itself more readily to the task.

Switch Traffic

The traffic capacity of a switching system is characterized by three related dimensions:

- Terminations
- Call attempts per hour
- System load in erlangs

The relationship of these parameters determines how a system is sized. This section considers the delicate balance between these parameters and what effect, if any, the introduction of data has on system sizing.

The number of terminations on a switching system normally is the sum of the customer's lines and the links (or trunks) to other locations. Some systems also include service circuits in the termination count.

The number of call attempts per hour is an indication of the capacity of the processor (computer) controlling the system. The processor is really a switching computer. That is, it does not perform a data processing function like a typical computer, but it is structured and operated like a computer.

A switched attempt consists of recognition of a call on the system, establishment of a connection through the network, and control of the associated signaling functions.

An allowance also is included for ineffective attempts, which are calls originating in but not connected through the network—for instance, when no digits are dialed. Anything the computer does in setting up or taking down a call should be included in this category. The number of call attempts per hour is measured from the originating side of the connection and can be determined by multiplying the average number of originating (including incoming) calls by the number of terminations.

System load capacity is a function of the number of connections through the network and the holding times of those connections, averaged over one hour. System load capacity should be measured in erlangs per hour and related to a probability of blocking. The probability of blocking may not be meaningful if the network is nonblocking, which occurs with some time-division systems.

The terminations, call attempts per hour, and system load in erlangs must be known not only for the initial system but also for any anticipated growth. This discussion emphasizes the call attempts and system load parameters, but the number of terminations is also important. The terminations often determine how far the system can grow and how easy these growth steps can be. Generally, easy growth should be possible up to two to three times the system's initial size.

An often-confusing characteristic of traffic is the method by which traffic intensity is measured. A typical system is shown in Figure 3.1. The system load consists of the originating (incoming) traffic plus the terminating (outgoing) traffic averaged over an hour. System load is normally referred to as mainframe traffic; some manufacturers also will show a switched traffic figure, which is 50% of the mainframe figure, to ease the confusion. In other words, the system assumes traffic intensity from your phone regardless of whether the traffic was originated or received. This relationship is true for other switches, because as the call is transported through the phone network the traffic intensity is measured at the incoming and terminating sides at every switch. This is illustrated in Figure 3.2.

Fig. 3.1. System Traffic Measurement Points

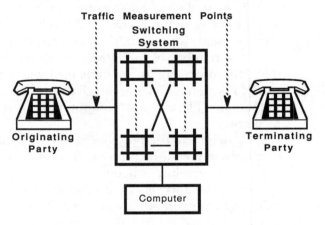

Traffic Parameters

Many traffic parameters should be familiar to the traffic engineer, including the following:

Number of sources   Normally refers to the number of items that can originate calls: an infinite number of sources can be assumed, which is valid whenever there are more than 25 to 30 sources.

Arrival rate   Refers to how calls arrive in the system: an average arrival rate should be used unless something unusual exists.

Number of servers   Must determine the number of switches, service circuits, links, and so on, required at the various stages of the system.

**Fig. 3.2.
Multisystem
Measurements**

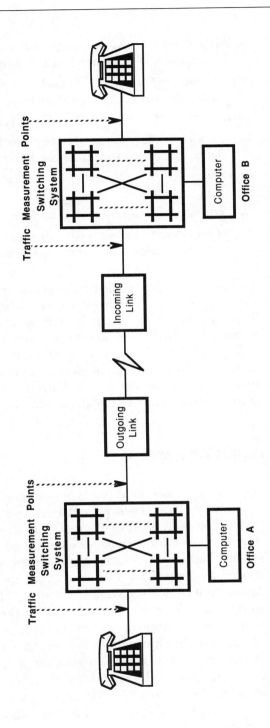

Holding time   Average amount of time a call holds a facility; a voice business call is approximately 2 minutes; data calls will vary.

Distribution of arrivals   Whether arrival times are exponentially distributed, which is a good assumption.

Distribution of service   Normally exponential.

Utilization   Determines whether the fraction busy or fraction idle of links, processors, and so on, is meeting good economic criteria.

Busy hour   The time of day when the demand is greatest for the various facilities.

Call count   The number of calls from the various sources.

Service criteria   Either lost call cleared (busy tone) or lost call held (dial-tone delay) at a specified grade of service.

The design of the traffic-carrying equipment is the means by which the various elements, links, service circuits, and so on, can be determined. These elements are normally arranged in groups to serve different customers. When all the elements in a group are busy, the call is blocked and will either be a lost call or a delayed call, depending on the group of elements. These groups of elements providing service are often referred to as *channels* or *servers*, and the devices that offer or receive calls—subscribers, terminals, or preceding switches—are *sources*.

Types of Congestion

A call can encounter three types of congestion or blocking as it attempts to go from one destination to another: internal congestion, network congestion, and line-busy congestion. Internal congestion is encountered during connection to a service circuit or a terminating source. The ideal system allows any source to talk to any other source at any time, as shown in Figure 3.3. However, such an ideal system is impractical because of its enormous size and cost. Modern technology can achieve something very close to ideal without the high cost. Digital switches are either nonblocking or nearly nonblocking: that is, there is almost no blocking within the switch. Nonblocking means the call can go through a multistage switch and be treated as if it were in an ideal switch. Electromechanical switches, which were used before digital switches, were rated on the amount of internal congestion or blocking they had. In other words, what is the probability that we cannot get from an originating source to a channel or terminating source. Switches had ratings like P.02, which meant that 2% of the time during the busy hour a connection could not be made through the switch.

Fig. 3.3. Ideal Switching System

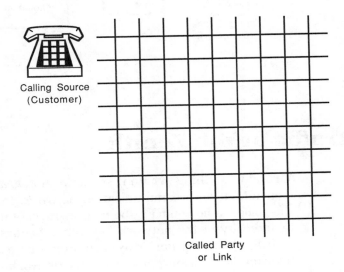

Calling Source
(Customer)

Called Party
or Link

Network or external congestion is the inability to obtain a channel (e.g., link) between switches. These groups of channels also are given blocking ratings, such as P.02, which now refers to all channels within a group being busy regardless of whether the call can get through the switch to seize the channel. The groups are sized so that the number of calls that fail due to insufficient quantities are predetermined and based on economics. The new translation techniques available with computer-like controls for telephone switches now allow multiple routing of a call toward its destination, thereby minimizing this congestion.

The third method of congestion is the blocking encountered when the terminating source or called line is busy, and this type is by far the largest of the three and the simplest to determine. If the line is engaged 10% of the time during the busy hour, then blocking should be encountered 10% of the time. This ignores any forwarding of the call to another phone or call waiting that would influence this number.

All three types of congestion will return a busy tone to the calling source or customer, although there is a difference in the cadence of the tones to inform the knowledgeable person of the type of congestion. This congestion or blocking is a function of the time of day when the call was made, although it is assumed all references are to the busy hour. Figure 3.4 illustrates the types of congestion that can be encountered.

Fig. 3.4. Delay or Loss in a Telephone Connection

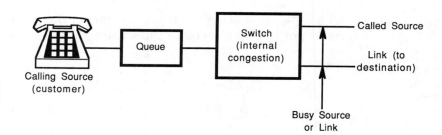

3.2 Traffic at the Zenith

Virtually anything written, spoken, photographed, or recorded can be transmitted on a communication network. In the vast majority of cases the voice traffic will have the most significant impact on the traffic volume or intensity of the network, but other factors could impact the number of calls the computer may encounter. For example, if electronic funds transfer is a major portion of the network, then it must be counted, because the number of calls would be critical.

It is therefore necessary to know which factors are candidates for the networks so that their number can be minimized. This list of relevant factors should include airline reservations, stock market quotations, hotel/motel reservations, and remote library browsing. Again, the amount of traffic intensity represented by voice communications should exceed the combined traffic in all other categories by a wide margin.

For our study we assume that the major components of the network are voice, data (e.g., electronic fund transfer), and facsimile. The next step is to project the amount of traffic these components will contribute to the network in terms of calls and traffic intensity.

Normally, traffic data is referred to the busy hour. The busy hour is the period selected to strike a balance between the uneconomical situation of providing equipment for every call and the unsatisfactory service condition in which only equipment for the average number of calls is provided. The distribution of traffic during the day in a typical office is shown in Figure 3.5.

Note that the volume is heavy from 9 to 11 A.M., falls during the lunch period, and rises again in the afternoon. A traffic study is required to determine from this distribution pattern exactly where the busy hour occurs. The traffic study should be taken in 15-minute or 30-minute increments. The International Telephone and Telegraph Consultative Committee (CCITT) defines the busy hour as the four consecutive 15-

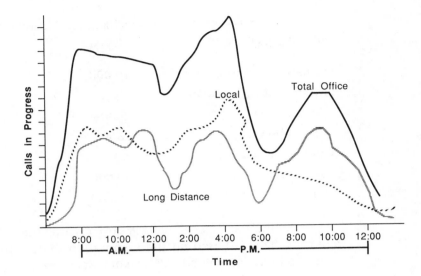

Fig. 3.5. Traffic Distribution during a Day

minute periods in which the traffic is the highest. Many offices use the two consecutive 30-minute periods with the highest traffic to determine the busy hour. Once the busy-hour period for a given switch has been determined, information from approximately 10 study periods must be evaluated to determine the average busy hour. Economics dictate that sufficient equipment be provided to meet a stringent grade of service during an average busy hour at a specified probability of blocking. These 10 study periods should be obtained during the busiest time of the business year; they probably will be Monday through Friday for two consecutive weeks.

Basic traffic can be defined in a number of ways, but it is generally agreed that the 10-high day (10HD) is useful for estimating switch and link requirements. For links between two nodes or offices, the traffic must be determined in both directions if a common set of links is shared. This is necessary only for the traffic intensity, not for call attempts.

Two traffic factors are important in the study: traffic intensity and number of calls. Traffic intensity, measured in erlangs, is used to determine blocking in the switch network and link requirements between offices. The number of calls is used to determine where the switch limitation exists for the computer or processor.

The traffic intensity uses the average of the 10HD, whereas the number of calls uses only the high-day figure. The high-day figure is also called the peak traffic, and it is common to see real-time capability expressed in average and peak figures; the former refers to the calls in the average busy hour, the latter to calls during the highest busy hour.

Consider a switch traffic specification that might be received from a manufacturer.

Number of terminations	10,000
Traffic load (erlangs) at P.01	
Mainframe	3000
Switched	1500
Number of calls	
Average	40,000
Peak	48,000

The number of peak calls is 20% greater than the number of average calls. This is based on the assumption that, with a normal distribution, the high day will be 20% greater than the average derived from the 10-day study. If only traffic volume is known, this assumption is useful because the average number of calls can be derived from the traffic load and the holding time. The formula for the traffic intensity is

$$\frac{\text{Number of calls} \times \text{Holding time (seconds)}}{\text{Traffic intensity in erlangs}} = 3600 \text{ seconds}$$

This formula can also be used to derive the number of calls, once the erlangs and the average holding time are known. For example, if a holding time of 180 s is assumed and the 1500 E of switch traffic is used, the formula would yield a call requirement (average) of 30,000. If a holding time of 120 s is assumed, the call requirement would be 45,000, which would exceed the capability of the computer during the busy hour.

3.3 Service versus Economics

A telephone switching system normally has a concentration stage as its initial stage, because individual lines have low occupancy and concentration offers the greatest economy. If we were designing the initial stage for, say, 60 customers, we might provide 24 serving devices (e.g., links to another stage) to ensure that all calls were handled. However, the structure of the equipment or the economics of the network may warrant only 20 serving devices, provided the grade of service is within our specifications. Therein lies the traffic engineer's function—economics versus service. Additional information regarding this problem is required;

for example, will the subscribers merely encounter delay if there are no devices or will the call be lost?

Another problem facing the traffic engineer is whether this initial stage is the only one or a part of a complex network, where all of the congestion specification in one area cannot be used. The service objective in this case is to attempt to provide the same loss criteria to a call across the network as to a call within the switch. This is not always possible in the current environment because the service criteria of a carrier are unknown. The basic answer to this dilemma is to attempt to reach the carrier with little or no loss.

Billing records should indicate how well the carriers are performing and over time will allow the customer to decide whether to change carriers.

3.4 Traffic Understanding and Administration

Background

In addition to its theoretical aspects, traffic engineering involves maintaining an on-line communication system, planning additions, cost improvements, new or expanded services, and other daily functions. Network administration ranges from line and link assignment for good balance to serving the network during unusual load conditions. Network administration depends quite heavily on setting service criteria and on good collection of information that will indicate whether these service criteria are being met.

The development of the service criteria and the use of the erlang are dealt with elsewhere. The amount of traffic or traffic data collection and validation need some additional discussion.

Traffic data collection is vital to almost every function the traffic engineer performs. Traffic data is necessary to size the various link groups, but it can also be used to determine the switching machine and associated computers. Remember, we said that the same formulas are used by the computer engineer as well as the traffic engineer. Thus, this function is extremely important because a great deal of capital will be committed based on the figures provided by the traffic engineer. Traffic data collection means determining the needs of the customers who will be serviced by this network, including voice and data requirements or any service that could impact the system and the network. Ideally, this traffic data can be collected from usage reports and projections of future needs. In practice, the information available, if any, is vague, and the traffic engineer will base most of the projections on opinions of usage or forecasted usage.

All this data, along with the efforts of other teletraffic studies, is likely to create an abundance of information and forms. Computer programs will assist in sorting the information, thereby permitting projections to be made properly.

Other Functions of Traffic

The traffic engineer is also responsible for the administration of lines, links, and switching machines from a capacity standpoint. Administering lines and their associated numbering are critical functions because every line will not select or receive every service or capability. Also, these services and capabilities must be reported to the switching machine and associated computers for implementation.

Administering the link group deals with determining if the traffic forecasted for a particular link group is being carried, and if it is not, finding out why. Link group servicing involves determining whether the link is actually in service and is concerned with the negotiations with the switch at the other end of the link group when service is interrupted. How well this function is handled depends on whether both ends of the link are owned by the same company or, if not, on how diplomatic the traffic engineer is.

The traffic engineer determines if the switching machines and computers are properly loaded and if a balance exists between the traffic functions.

The traffic engineer must also determine when a central office needs relief. This relief can be in the form of additional lines, links, or service circuits, or, in extreme cases, the office itself. The traffic engineer is responsible for the linking between the offices within the network.

The basic questions are: "How is the amount of traffic in an office determined?" "How is the amount of traffic on a link group determined?" These questions may sound basic, but their answers are very important to the basic switching structure and for determining communities of interest within the network.

To perform all of these functions properly, the traffic engineer must determine the traffic parameters about which information is needed, the design of the traffic-carrying equipment used, and the preparation of the traffic orders.

3.5 Conducting a Traffic Study

Most articles on network planning do very well when they talk about setting objectives and laying out a network, but they fall from grace when they talk about traffic studies. Normally, they say something like, "take a traffic study and use the results for the plan." Since most readers bought the book to help them make a traffic study, this statement is most disappointing.

The basic function of the network is to carry traffic, and an essential ingredient of any communication plan is determining the traffic between nodes (switching points) and making plans based on this distribution. For our purposes the traffic study will be divided into the following increments, each one using information from the previous increment: network plan, cluster plan, and exchange plan. These three areas are illustrated in Figure 3.6.

Fig. 3.6.
Network
Configuration

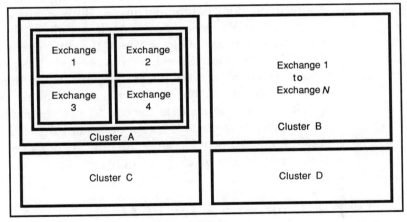

3.6 Switch Calculation

The switch (or exchange or node) equipment is a basic facility that must grow in discrete increments to meet the customer's demands. Because many of the exchanges will have similar characteristics, it is convenient to have a program or electronic spreadsheet that will enable the planner to determine the requirements for a particular switch and what equipment will best satisfy these demands at the least price. The program or spreadsheet must be able to

- Evaluate the alternatives for an exchange
- Evaluate the growth capability of the alternatives
- Estimate the required equipment for the busy season
- Determine whether the selected arrangement can meet the service criteria for the exchange
- Estimate the cost of the exchange and the units of equipment required

Strictly speaking, the traffic engineer should use the program or spreadsheet to determine the equipment for the switch, and this decision can be independent of the cluster or network. However, there are certain advantages to having the switches handled by the same or similar equipment, especially in the administration area. The most difficult objective to place in a program is the service criteria, because manufacturers have different methods for calculating their service circuits and other traffic-dependent equipment items. However, digital switches have narrowed these differences, so most, if not all, of the equipment should be identical to our calculations. If not, the differences should be explainable.

The information needed from the switch is primarily the number of lines and trunks and the traffic-carrying requirements of these units. The manufacturers will have to supply the breaking points for the building blocks and any other data needed to size the exchange. Some manufacturers may be reluctant to supply detailed price information, but they may be able to provide an estimating formula or, given the parameters they need, give a quick price quote.

Traffic Characteristics of Lines

The single largest item for a switch or exchange is the traffic characteristics of the lines. They determine not only the equipment needed to handle the lines but also the traffic on the links. This information together with their destinations determines the link terminations on the switch. Almost all terminations for the switch will be lines and links: the manufacturer should be able to indicate any additional terminations necessary.

Let's determine the line traffic. Lacking any additional information, it is safe for us to assume that, on the average, each customer will use the phone 10% (or six minutes) of the busy hour for voice traffic. This usage figure can be expected to increase at the rate of approximately 2% per year, which is phenomenal for a 100-year-old commodity and apparently indicates continued growth for this industry. For the real-time processor or computer controlling the system, we'll say the six minutes of occupancy consist of two calls, each lasting three minutes, with one call originating and one call terminating. Ineffective attempts associated with originating attempts are assumed to be part of the sizing for the processors.

Data traffic does not have 100 years of information available and is more difficult to estimate. Also, data traffic currently tends to consist of extremely long sessions or exceptionally short inquiry-and-response calls. With the introduction of workstations, which can download information from the main computer and process the information without being connected to the computer, the long sessions will be reduced.

Therefore, we emphasize the inquiry-and-response type of data traffic. For the study we assume two inquiries and responses, each consisting of 100 characters to the computer and 600 characters from the computer.

For facsimile we assume that one page of printed material consisting of 55 lines of 70 characters each will be transmitted during the busy hour, a fairly typical page. Eighty characters per line is possible, but some space must be provided for margins. Video is assumed to be handled via a special terminal and system and will not be included in our calculations.

The traffic per line should be brought to a common base before the calculations are performed and total traffic computed. The common base for these services is bits per hour, and the conversion method is

- voice = 64,000 bits per second (b/s)
- data = 10 bits per character
- facsimile = 10 bits per character

The bits per character may vary slightly, depending on the method employed, but the figures are conservative and should represent the worse case.

Line Calculation for the Network

A modern phone or workstation capable of handling voice, data, and facsimile is equipped on a digital system under study. The workstation operates with only one connection to the switching system. This assumption is necessary because many workstations have two connections to the switching system: one for voice and one for data. This arrangement changes the characteristics of the traffic from the line to the system.

The traffic for the workstation under these assumptions is shown in Table 3.1.

Table 3.1.
Workstation Calculations

Voice calls × holding time (sec)/call × b/s	$2 \times 180 \times 64{,}000 =$	23,040,000
Data inquiry/response × characters/inquiry × bits/character	$2 \times 700 \, (100 + 600) \times 10 =$	14,000
Facsimile pages × lines/page × characters/line × bits/character	$1 \times 55 \times 70 \times 10 =$	38,500
Total		23,092,500

This example illustrates why voice traffic is important on a network load and why it is reasonable to assume that data traffic, which has only occasional use, can be added without a serious impact.

Another method of calculating the traffic uses the erlang because it deals with occupancy, which also can be obtained from the foregoing information. This is especially useful in situations where the data speed has not reached 64,000 b/s through the network.

The line occupancy was given as 10% or 0.1 E. The data and the facsimile speed would have to be known; for this example assume 1200 b/s. There are 52,500 (14,000 + 38,500) bits, and this equates to 43.75 s of occupancy (52,500/1200). This represents an occupancy during the hour of 0.012 E (43.75/3600 s). Add the two figures (0.1 for voice and 0.012 for data) to obtain the total occupancy of 0.112 E. Again, the greatest contributor is the voice traffic; as the switching speed for data increases, the data occupancy will decrease for the same calls. Once the data speed reaches 64,000 b/s, the figures will be equivalent to Table 3.1.

Impact on the Computer

The voice traffic consists of one originating call and one terminating call. If the terminating call was from a distant office, the processor would have handled it as an incoming call and the call would have been counted in calculating the capability of the processor. Calls that originate and terminate within the same office are counted only as originating calls. (Remember, the processor capacity is determined by the number of originating and incoming calls on the system.) For our example assume that all terminating calls were from incoming requests and are included in the count.

System specifications indicate a capability of 40,000 calls average; if each termination yields two calls, the switching system could grow to 20,000 terminations if other parameters permit it. However, when the data and facsimile traffic are added to the voice traffic, the number of calls per termination increases to 5 (2 voice, 2 data, and 1 facsimile). A switching system with this configuration and a processor capable of 40,000 calls can be equipped with 8000 terminations.

Nonvoice traffic has its greatest impact on the call-carrying capacity of a switching system and little impact on the call volume or call intensity portion of the system. There can be exceptions to this generalization, depending on the application.

If a telecommunications manager is studying the impact on a switching system of adding data and other nonvoice functions, call-carrying capacity is a useful figure to know. Usually, such studies focus on the impact on traffic intensity, but usually the impact in this area is minimal. The processor is the component that normally must be improved. There may be, in some special application, a PBX that switches virtually full-period terminals to a distant mainframe computer. This type of application is related totally to traffic intensity and would overload the network but not the processor, but this would be an anomaly in telephony.

The typical application of the system is to enhance the productivity of the users by providing voice, data, and facsimile from the same terminal. For such applications we should study the processor to determine if there is sufficient capacity to perform the required tasks.

With voice traffic accounting for the vast majority of workstation traffic, it is tempting to base the service criteria only on voice traffic. Service criteria, however, must be based on the requirements of all the services, and each service must be treated separately to ensure that proper levels are being met. Voice traffic service criteria is in dial-tone delay and post-dial cut-through, whereas data traffic criteria is normally in terms of response time. Response time is the interval between the last keyboard entry and the display of the first character from the computer.

The typical service criteria for a computer not connected to a telecommunication network are given in statements such as: "90% of the inquiry must not exceed 6 seconds." The telecommunication system typically is judged on its total performance and not on the criteria of the computer. When the telecommunication network is added, its service criteria will have a statement like: "99% of the call will be switched within 1 second." That is, the terminal is connected to the computer within that time, and now the computer response time is added to that figure. These figures are additive, but any difference caused by the telecommunication network should not be noticeable to the user.

This example also illustrates the need for a unit of measure for many bits. The example is shown for one workstation: the traffic through the network may consist of 1000 workstations or 10,000 workstations, and the number becomes unwieldy. A term similar to the term for space distances is needed. We can speak of distances between stars in light years, but if we had to speak of those distances in miles we would have great difficulty with the size of the number. The same is true with bits per second. If telecommunication is to use bits per second, a new term is needed to simplify the calculations.

Calculations for a Remote

The preceding example clearly indicates that, at least for now, data traffic will not exceed voice in typical applications unless the network is devoted to a special data function. This is true even if all the data functions are implemented on the network with the exception of video. The introduction of video will severely change the network, but the growth in this area awaits the wiring of offices and customers' premises with fiber optics.

People have become accustomed to a network being supplied by AT&T and have taken for granted the many aspects of these services. A new generation of communication specialists eventually will dominate the industry, and they will not use AT&T's techniques because their goals will be different. The communication specialist will be concerned with communicating with computers as easily as with people. They will also be concerned with transmission links that will carry the new technology and not with the idea of trying to push 10 pounds of data in a 1-pound sack. They will look for a universal terminal that will be inexpensive and yet serve the most demanding requirements—a telephone of the eighties.

When the network for this communication system is constructed, it will start with the remote unit, determine its requirements, and proceed to build the interfaces. It is assumed that the remote unit will be a local area network (LAN), which will handle voice/data, or a concentrator. A concentrator can be a remote-switching unit, a multiplexer, or any other special voice/data unit. The requirements of the remote and calling within the remote, if needed, will determine which unit is used.

The easiest answer to the communication link between the remote and the switching systems is a fiber-optic link capable of meeting the traffic requirements of the information being transmitted. In some cases it may not be possible to install a fiber-optic link (e.g., due to right-of-ways or terrain difficulties). If so, then a microwave can be used. Several small rooftop-type antennas can operate at the T1 or T2 carrier rate (1.544 or 6.312 Mb/s) and handle most remote applications. These systems operate at approximately 40 GHz and have a short range (around 10 miles), but are inexpensive and easily installed. They normally operate at a low error rate and will carry both voice and data in a digital format.

The formulation of the remote units with the telecommunication system is extremely important, because the method for switching the call within the system is somewhat determined by the number of remotes and the traffic on the remotes. Again, video is a major criterion of these remotes; if video is required, the structure of the system must be examined against a video background. It is true that many areas have restricted use for video during the day, but, for security, they would like to hook up the telecommunication system to TV monitors at night without using individuals at these remote offices. This is especially true on a college campus or an industry complex where several buildings are interconnected by the communication system. Compressed video can use the T1 rate of 1.544 Mb/s if arrangements are made to switch from voice/data to video. We can begin to see the need for planning the total network against all the requirements to achieve the full benefits and cost advantages of a communication system.

Identifying the traffic requirements for a remote unit is normally difficult. Usually the unit will be a new installation with no knowledge of the traffic flow, or it may exist under the auspices of the local telephone company and data may be unattainable. If it was controlled by a telephone company, they should share the information with us. However, they may not have taken a study for a long time. Let's assume our remote is manned by 35 people, and we would like to install a T1 line (1.544 Mb/s) so that at night we can connect the unit to a TV monitor. The traffic from the customer lines is as follows:

- Three telephone calls per line at 180 seconds per call
- Two inquiries/responses per line
- One page of facsimile per line

The traffic calculations again are based on 64,000 b/s for the phone calls, 7000 bits (700 characters) for each inquiry/response, and 38,500 bits for a facsimile page.

From Table 3.2 the total traffic is 1.21 Gigabits/hour (Gb/h) offered to a T1 line. The T1 line capacity is 1.544 Mb/s × 60 s/min × 60 min/h or a total of 5.558 Gb/h. To obtain the occupancy on the T1 line, divide the usage by the available usage: 1.21/5.558 = 21.8%. A standard traffic table would show that the T1 line could handle the foregoing traffic at an excellent grade of service.

**Table 3.2.
Remote
Calculations**

Per-line calculations	
$3 \times 180 \times 64{,}000 =$	34,560,000
$2 \times 7000 =$	14,000
$1 \times 38{,}500 =$	38,500
Total	34,612,500
Total $\times$ 35 $=$	1,211,437,500

These calculations are based on the premise that the customers in the remote unit have full access to each of the 24 channels of the T1 line. This is a very important assumption, one that is normally true in the digital world. The table is based on an erlang's lost call cleared formula, which states that if a call encounters congestion the call disappears from the system never to appear again. This assumption may not be totally true but it works very well in reality when we assume an infinite number of sources, and we know the sources are finite. When a finite number of sources is sending traffic to a switch, the ability of the sources to generate traffic is diminished as they occupy channels. As an extreme example, if 24 customers were accessing 24 channels, the 24 channels could have 100% occupancy. That is, the 24 customers could never overload the channels.

These traffic figures apply whether the remote unit has phones, time-sharing options (TSOs), or workstations as long as all the units are attempting to access the various channels in an orderly manner and the total traffic is as shown in the table. Also, if the remote has some special characteristic (batch, printing, etc.) using the T1 line, this traffic should be included and broken down into bits per hour as a common denominator.

3.7 Developing Clusters

The introduction of digital switching has made the traffic administration of switching systems easier through nonblocking networks. It has not changed the administration of links between switches. If anything, it is now more difficult with the introduction of concepts such as most economical routing and bypass. The basic function of link administration remains the same—having facilities available when they are needed.

In addition to the link forecast and administration, special service circuits associated with links represent a sufficient part of a switching system and must be properly sized and commissioned. Normally, links and service circuits occupy the same mounting space within equipment frames. Proper care must be taken to ensure sufficient space and frames for these units.

An added dimension to the current environment is the requirement to maintain a current inventory of various routing schemes and carriers so that periodic investigation of cost for the routing strategy can be made. This cost is constantly undergoing change as more carriers enter the marketplace and the competition accelerates. This function may not fall directly to the traffic administrator, but, at the least, the ability to make rapid changes based on new cost figures must be included in the duties.

The initial approach to the link group forecast is to define the switching systems or nodes that will belong to a particular cluster. The method of linking and handling overflow traffic will, most likely, be different for traffic within the cluster versus traffic outside the cluster. The switching systems within the cluster should be determined by geographical area and community of interest. The latter is based on the amount of traffic between particular offices and should be known when the study starts. For example, if there are 40 E of traffic between office A and office B, 2 E between office A and office C, and 1.5 E between office B and office C, then A and B belong in the same cluster and C probably doesn't belong to that cluster.

Once the cluster is determined (and changes to it are encouraged as the traffic patterns shift over time), the traffic between all the nodes within the cluster should be gathered. Many individuals doing link forecast will do both a short-term forecast (less than 5 years) and a long-term forecast (5–20 years). At this time it is sufficient to do a short-term forecast; once this is validated, we can think about a long-term forecast. The latter is useful for deciding where new nodes will be located.

The linking arrangement between the nodes or switches within the cluster is complicated and subject to iterations before the minimum cost at an objective grade of service is achieved. It is highly recommended that a computer program be developed or purchased for this task. The program should include or be able to

- Derive the projection of future loads based on point-to-point input and annual growth
- Determine the point for justifying direct-link groups based on traffic and facility cost

- Show first-route load and overflow load on groups carrying alternate-routed traffic
- Calculate economic point for direct-routed traffic versus overflow to alternate
- Determine economic point for direct (or bypass) linking as opposed to using a common carrier
- Calculate peakedness factors for overflow traffic and show equivalent links
- Show where digital carrier can be justified
- Compare cost of all direct linking versus minimum arrangement

The last item is needed to compare a fully direct-linked arrangement with the optimum because the former is much easier to administer and can handle unusual overload without affecting the total network. Translation within the switching system is easier with direct-routed linking. However, the typical cost differential for most applications will favor alternate routing.

A recommended arrangement for a cluster is illustrated in Figure 3.7.

The traffic administrator should understand the program and be able to vary the inputs, depending on the particular requirements within the network. There should be traffic figures for switch traffic, traffic between switches in the cluster, traffic to the public network from each switch, and total traffic to the rest of the network and the long-distance carriers. The program can be repeated for each cluster, which really demonstrates the value of this approach.

There will probably not be a pure tandem node in a cluster ("pure tandem" typically requires a large network), but certain nodes provide tandem functions for other nodes and allow traffic from several sources to be combined to justify a link group or a digital carrier group. The latter is very valuable in a digital environment because these units will eventually permit a call to be transported across the network in a digital format.

3.8 Network Determination

Once the various clusters have been determined, the figures should include the traffic to other clusters within the network. If not, these numbers must be somehow ascertained before the network calculations

**Fig. 3.7.
Arrangement
for a Cluster**

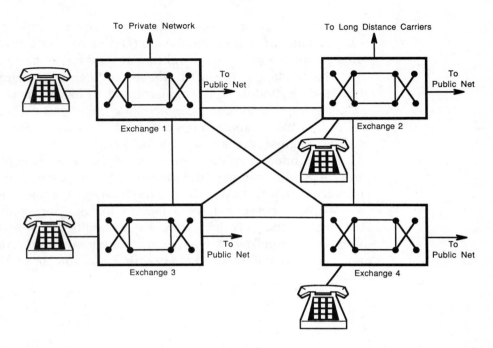

are started. A lot of networks get built through a reaction mode; that is, if someone insists on a node at a certain location or on a link group to a certain location, the node or link group is provided and the network evolves in this manner. Certain nodes and link groups will emerge within a network in this manner, and we must simply take a pragmatic view of them and continue to evolve the network based on good traffic practices.

Besides the political aspects of the network, the traffic distribution, the type of network, and the need to stay digital throughout the entire call are very tangible problems that can be solved. The type of network will be discussed in other sections, but it can have a large impact on the structure.

It is probably possible to use the program developed for the clusters to do some of the network optimization, although the overlay of the digital facilities with the existing analog ones presents a unique problem that may not be handled by this program. Whether the network requires another program or whether we can optimize it by hand depends on the size of the network. If the network contains four or more large clusters, we should consider a program. If it contains less, we might try to optimize by hand. In either case it is an iterative process subject to a number of variables, including political.

Linking between clusters is similar to linking from the nodes or switches within the clusters. The main difference in the network is the driving force to commission digital and replace analog. For large networks a tandem switch may be justified as well as a node that handles some customers and a large volume of tandem traffic.

The tandem is the area where modular link engineering can be employed. In modular link engineering the links are arranged for 12- or 24-link groups, a common denominator for digital carrier groups. This will permit faster employment of digital. If a computer program is used, modular engineering is easy to implement.

The tandem location is excellent for the network control center. The physical administration of the various link groups can now be centralized and permit this and other functions to have a common home.

We have presented some factors to consider when deciding the structure of the network. They are in addition to the calculations of direct links, overflow, bypass versus common carrier, and other necessary decisions within the network that are similar to those made for clusters.

4

Transmission

We finally come to an area where everyone has a pronounced judgment. There may be no universal agreement on whether a connection is providing good transmission; nevertheless everyone, from the least to the most knowledgeable, will express his or her opinion on transmission. We currently live in a world of two transmission systems, digital and analog, and some understanding of both is needed to appreciate what goes into this aspect of the network.

4.1 Information

Information can take a variety of forms: pictures, sound, taste, feel, and smell—in other words, any of the senses can receive information. We also transmit information, normally through sound, although body language and other nuances can convey information. How then is information defined? A highly respected definition is the following: information is that which is capable of clarifying uncertainty in a given situation. I prefer "a signal that conveys intelligence." In either case, to transport information, according to Claude Shannon, an arrangement similar to that shown in Figure 4.1 is necessary.

The transmitter is the information source where the information is created. It also serves as the input for the message to the system. The sound waves from a voice or the keystrokes from a typewriter are examples of transmitters.

The encoder is used for the processing and storage (if necessary) of the message. For example, it is responsible for converting voice waves to a digital format in digital networks. The encoder is a very complex part

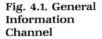

Fig. 4.1. General Information Channel

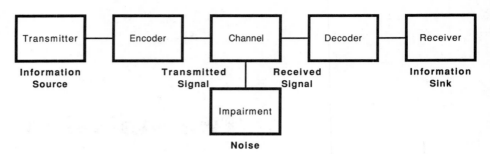

of information theory, because it is required to perform tasks involving reorganization, compilation, calculation, and consolidation of the information. The end result of this process is a transmitted signal toward the destination.

The channel is the transmission media for the distribution of this information. The media is air if two people are conversing in a room, but it takes on more complex forms as the information is transported across distance. The channel will usually operate with the information, but it also may pick up impairments (noise, transmission loss, echo, or anything that will interfere with good conversation or data transmittal) detrimental to the information. A major concern of the channel is the understanding of these impairments and their elimination or reduction. Noise, called *gaussian noise*, is the largest impairment to the transmission of good voice and data information.

The decoder is responsible for converting the encoded information back to an intelligent message for the receiver.

The receiver is the party at the other end of the connection or, technically, the information sink. The receiver can be the ear, a voice storage memory, a mainframe, or anything capable of understanding or storing information.

Information theory has become critical in the digital era, because all information is coded to its simplest level, 1's and 0's, before transmission. Both the telecommunication industry and the computer industry are convinced that each is the best coder and handler of this information. This area is where the two disciplines clash with different coding methods and techniques.

In addition to defining the way information is transported, information theory also provides codes for the way messages are expressed. The information code can be broken down as follows:

- Elements: the smallest building block of a code—in digital format the bit is the element of the message

- Characters: made up from elements
- Words: consist of characters
- Messages: groups of words make up a message

Codes can take different forms (for example, ASCII, EBCDIC) at the elementary level. The structure of information theory, therefore, provides the fundamental units that must be compatible if information is to be transported with intelligence across the room or across the country.

Transmission is concerned not only with the channel or the transmission media but also with the interface between the encoder/decoder and the channel.

4.2 Transmission Media

Switching systems or serving offices greatly reduce the number of wires necessary to interconnect a community. They also divide the field of transmission into two distinct areas: the connection from the customer's premise to the serving office, and the connection between the various serving offices. The former has remained virtually unchanged for the last 100 years, but is about to be transformed as fiber optics replaces copper wire.

The interoffice connection has undergone numerous changes as scientists and technologists developed methods to obtain faster, more universal transmission and to place as many conversations as possible on one wire between two serving offices. The transmission area early recognized the value of digital transmission and its inherent ability to carry 24 conversations on two wires.

However, before we cover the various transmission media, we give a little background on transmitted information. Any transmitted information, such as voice, data, light, or radio, can be defined in terms of frequencies. Frequencies can be transmitted through media: the higher the frequency, the faster the transmission speed. That is, an electrical signal traveling along a transmission line takes time to get from one point to another. The amount of time is related to the frequency: as frequency increases, time decreases.

The basic information can be transmitted in analog or digital form on virtually all transmission media; however different techniques must be used, depending on whether it is analog or digital. Analog uses amplifiers to strengthen the signal over distance, whereas digital uses repeaters.

Early in its development, the telephone company found that it did not have to transmit the full range of the human voice. It also noted that the quality of a voice channel can be measured by two parameters: intensity (energy level) and intelligibility (speech clarity). Most speech energy is concentrated in the lower range, whereas higher frequencies contribute most to intelligibility. Based on this and economics, the majority of telecommunication techniques transmit between 200 and 3200 Hz.

The transmitted range of frequencies from 200 to 3200 Hz is known as the bandwidth. Because the human voice has a wide range of frequencies that can travel at different speeds, delays tend to distort the conversation. This is why transmission engineers are paranoid about delays in circuits or systems.

The frequency of an item refers to the number of cycles per second it occupies in the spectrum. The measurement for frequency is cycles per second (c/s) or hertz (Hz). For example, hearing occurs from approximately 20 to 20,000 c/s (20 kc/s) or from 20 to 20,000 Hz (20 kHz). The eye sees from 3×10^{14} to 6×10^{14} Hz.

As the frequency increases, the wavelength of the cycle decreases. The relationship is wavelength = V/frequency where V is speed of light. The spectrum is shown in Figure 4.2 with a few examples of its application. In AM and FM radio, for example, FM is transmitted at a higher frequency, but each station is also given a greater bandwidth and the quality of the sound is improved substantially with this method.

Decibels

Because most of the energy associated with speech is concentrated in the range transmitted, good recognition can be obtained over a narrow range. This energy or power of the speech is measured in decibels (dB). In telecommunications the decibel is used to measure the loss in signal strength, in addition to numerous items, and this has caused much confusion over the ages. Simply put, when a signal is transmitted over distance, it loses some of its strength; we need to know how much is lost, so we can add that amount back. The unit used for this measurement is the decibel. For example, if we know the signal loss is 1 dB/mile and we are transmitting 20 miles, then we know the signal loss is 20 dB at the receiving end; therefore we need to amplify the signal by 20 dB. The decibel can also be quickly converted to the amount of power that survives when the signal is received, which is also a convenient unit to know.

Fig. 4.2.
Frequency
Spectrum

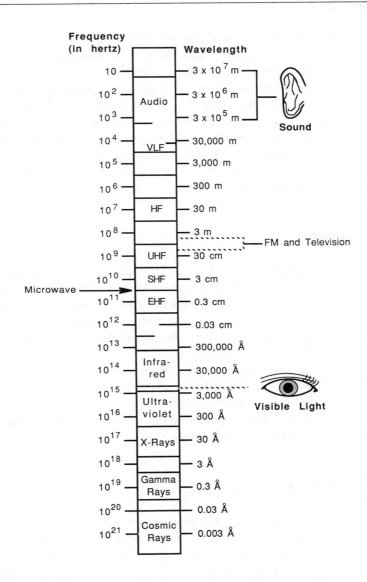

Bandwidth

As previously stated, the bandwidth of hearing is from 20 to 20,000 Hz, although a telephone channel is only required to transmit a portion of that range—from 200 to approximately 3200 Hz—to ensure a reasonable conversation. The channel is therefore said to have a bandwidth requirement of 3200 Hz. Facilities to handle this bandwidth or higher can be used to transmit the information. In an AM radio station the carrier frequency assignments are spaced 20 kHz apart. Each station is allowed to transmit power on the center 10 kHz of this range. This provides a 10-kHz guard band between adjacent stations. FM stations have a bandwidth of 18 kHz and are therefore capable of transmitting a much greater voice range.

Of the various transmission units, bandwidth will take on greater and greater importance because it is as important in digital transmission as it is in analog. It also is critical in the area of Picturephone or video, because it is directly related to the quality of the picture. As fiber-optic units become more and more common, the bandwidth problem will begin to be solved.

Cable

Paired cable is the most popular transmission media. It usually consists of a group of copper wires twisted into pairs. These groups are normally 25, 50, or 100 wires, but other arrangements are possible. Copper pairs are employed typically between the switching office and the customer's location, although they also are used for links between offices where the distance is short.

For long-distance circuits coaxial cable is used because its transmission characteristics are better than those of twisted pair cable and it is better protected from the environment. Until microwave radio coaxial cable was the principal communication method for long-haul traffic. Coaxial cable is used extensively in cable television and data communication.

Microwave Radio

The main advantage of radio over wire transmission is the absence of physical facilities between two points of communication. Certain sections of the spectrum chart have been reserved for radio transmission, both terrestrial and satellite, in the 1–30-GHz (gigahertz) range. In addition, frequencies just below 1 GHz are reserved for cellular mobile radio (CMR). The use of radio for communication depends on line of sight between transmitter and receiver and can provide relatively constant loss with frequency over wide bandwidths.

In microwave (radio) transmission the information received depends not only on the power of the information transmitted but also on the characteristics of the transmitting and receiving antennas. A good terrestrial transmission system requires a large wavelength, an efficient antenna, a line-of-sight operation, and a large receiving antenna. Antennas will often transmit and receive at the same time. From a system stand-point terrestrial microwave consists of terminal stations, switching stations, and repeater stations. An arrangement is shown in Figure 4.3.

**Fig. 4.3.
Terrestrial
Microwave
System**

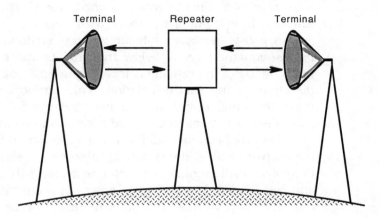

The terminal station is where messages are put on or removed from the system, a switching station provides a place for routing or rerouting groups, and a repeater station reamplifies the microwave signal every 30 miles.

Atmospheric conditions severely affect the received signal level in terrestrial microwave systems. Various schemes are used to compensate for these conditions.

Terrestrial systems can be either analog or digital, although analog had a tremendous advantage over digital for years due to its efficient use of bandwidth. Recently, thanks to complex schemes and lower costs, the digital microwave is overtaking the analog systems and should dominate in the near future, as more networks attempt to stay digital from end to end.

Terrestrial microwave is better adapted than cable to spanning natural barriers, such as water and mountains, and unnatural barriers, such as highway right-of-way. Also, radio is efficient in bypassing the local telephone company; to assist this effort, frequencies have been reserved for business application.

Satellite Communication

Arthur Clarke first predicted satellite communications after WWII, but not until Bell Labs launched *Telstar* in 1965 did satellite communication become a reality. Unfortunately, the U.S. government decided that satellite communication would not be given to the Bell System, although the logic of that decision is elusive.

The era of the geosynchronous satellite was started in 1965 when the *Early Bird* satellite was launched. Today, space shuttles are placing satellites in orbit or fixing them with ease.

A geosynchronous satellite appears stationary to someone standing on the equator directly below the satellite and capable of seeing 22,300 miles. Actually, the satellite is traveling at 6900 miles per hour in the same direction as the earth's rotation. This positioning, speed, and distance make the satellite appear stationary relative to a point on earth, which allows earth stations to use fixed antennas to communicate with satellites.

Geosynchronous satellites are relay stations for radio signals between two earth stations just as microwave relay station towers are for terrestrial radio signals. The satellite receives the signal at one frequency and retransmits it with higher power at a different frequency. The obvious advantage of this type of communication is that a voice connection across 3000 miles of the United States would require approximately 100 microwave towers (one per 30 miles), whereas the same conversation using a geosynchronous satellite would be arranged as in Figure 4.4.

The conversation would be completed in one hop, but, ironically, the delays would be greater with the satellite connection. Both connections are transmitting at about 90% of the speed of light, but the satellite connection must travel 22,000 to 26,000 miles up and the same distance down, whereas the terrestrial connection only travels 3000 miles. The satellite delay is approximately 1/4 s in each direction and can be annoying to the customer, especially in the area of echo. Echo is the speaker's own voice transmitted back by equipment at the remote end. Typically, echo is expected by the speaker as part of the conversation. However, echo delayed by time can be distracting, so satellites use echo cancelers to eliminate this problem.

The bandwidths from the earth stations to the satellites are shared by either a frequency-division multiplexing (FDM) method or a time-division multiplex access (TDMA). The FDM method assigns different portions of the frequency band to different earth stations, similar to microwave techniques or FM radio. The TDMA method is similar to the PCM (pulse code modulation) technique, where each point is given access to the satellite for a few microseconds to transmit its information. TDMA lends itself to digital electronics and is commonly used in earth stations.

One use of satellite communication is to broadcast international events from some part of the world to anywhere else in the world. In addition, almost all videoconferencing is performed via satellite. In videoconferencing the bandwidth requirements of the connection necessitate

Fig. 4.4. Satellite Communication

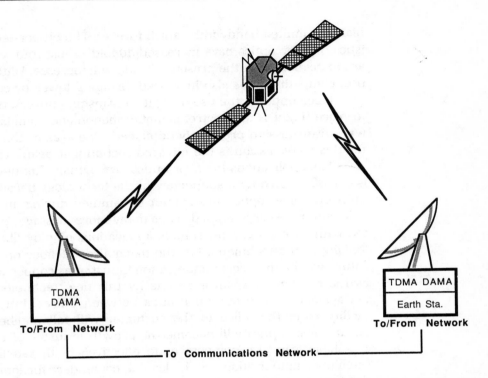

TDMA
DAMA

To/From   Network

TDMA DAMA

Earth Sta.

To/From   Network

To Communications Network

bypass of the present network, with private facilities being leased for the occasional use.

In areas where it is difficult to implement electronic mail or computer conferencing or other services, an inexpensive communication satellite in a lower orbit than the geosynchronous orbit is contemplated. In these satellites switching and information storage will be used to facilitate the services. The satellites will be placed in orbit so that an earth station can always communicate with one of them, though not necessarily the same one. The message will be transported to the current satellite, stored, and then delivered to the appropriate earth station connected to the destination point.

4.3 Fiber Optics

Introduction

The epitome of the changes taking place in telecommunication is fiber optics, the brightest transmission star of the new era. The extensive use of fiber optics in communication will make telecommunication engineers' fear of a limited bandwidth vanish forever. The bit-per-second character-istics of fiber optics have increased 10-fold in just four years, and there are indications that the present bit rate will increase. With this increased use, competition has also increased, bringing lower prices.

Fiber optics is the use of light to transmit conversations from place to place. Light is an electromagnetic phenomenon similar to the micro-wave transmission previously described. The area of the spectrum used for this transmission is the infrared section just below visible light.

The applications for fiber optics are virtually limitless. The military is one of the strongest supporters of the technology because communica-tions over fiber-optic links cannot be jammed during an attack and the link cannot be easily tapped. Thus many more messages can be transmit-ted without coding. Most robotic applications employ fiber optics or are looking at the technique. But the major user of fiber optics in the near future will be the telecommunication industry, as copper wire is replaced as the principal transmission line by this new technology. Initially the replacement will occur on groups between offices, but eventually the facility from the office to the customer will fall to fibers. When this occurs, fiber optics will become the growth industry of the decade.

Fiber can also compete quite effectively with satellites. Fiber and satellite communication are the leading contenders for long-haul traffic in the future. For both voice and data operations, satellites have exhibited problems that a fiber-optic link does not have. For example, there is an inherent delay due to distance on satellite calls, which initially produced a severe echo. This problem was corrected by an echo-canceling circuit, but it made the connection virtually two-wire—that is, conversation could not be interrupted. This delay and echo-canceler also interfered with error-checking techniques on data calls. Because of this problem, very little data traffic is carried over satellites. Various plans are being implemented to overcome this data transmission problem, but the vast majority of data transmission uses, and probably will continue to use, terrestrial circuits.

Another area of telecommunication for which fiber optics is being considered is the local area network (LAN) applications, because the growth or capacity of these switches is determined by the speed of the transmission media. Also, LANs require bandwidth for services such as high-speed data or video. Fiber optics provides a satisfactory answer to both of these problems and allows LANs to operate over a greater area than with coaxial cable systems.

The use of fiber optics for telecommunication was developed in the late 1960s and early 1970s, and even in those days the loss over a kilome-ter was less than 20 dB. Then why did it take so long for it to be used? Some reasons are (1) the battle over single mode versus multimode (single mode scored an "overnight" victory in the early 1980s); (2) splicing the cables when a fault occurred was extremely difficult until recent advancements; and (3) an understanding of and acceptance of the technology were almost nonexistent until recently.

For telecommunication the operation of an optical fiber or waveguide is similar to other communication methods. The electrical signal is converted to an optoelectronic (light) signal and passed through a cable that has characteristics similar to a waveguide. Sending this light or the on/off signaling of the message is performed by a laser (other devices are sometimes used), which can operate sufficiently fast to transmit the binary information.

Different speeds are available with fiber optics, but the de facto standard is becoming 90 Mb/s (90,000,000 bits per second), which is equal to 1344 voice connections and is normally composed of two 45-Mb/s units, the DS3 level of switching (44.736 Mb/s).

The ability of fiber optics and satellites to provide bandwidth previously considered extremely expensive has led many experts to conclude that there will be a glut of bandwidth on the market within a few years. However, the potential applications for this new bandwidth are not known at this time because the users of this new technology still live in a world of 1200 b/s. A tremendous educational program is necessary before it is known how rapidly the user can move from the 1200-b/s world to 64,000 b/s and beyond. Historically, all predictions of telecommunications glut on the marketplace have proved false.

Principles of Fiber Optics

Fiber-optic systems are presently considered closed light-wave systems; that is, they do not employ free space as the transmission medium: an optical fiber is used. Light energy propagates through the fiber to its destination, where it is detected. Light systems can employ free space, and these free-space systems are being investigated for certain forms of communications, especially between the instrument and the in-house wire, or for space communication. Here we ignore free-space systems.

To illustrate how a fiber operates, see Figure 4.5. A light source is illuminating the end of a short section of pipe. The pipe wall is assumed to be highly polished, so the light rays are reflected many times as they traverse the distance. If the pipe is relatively wide, the rays will be reflected many times, resulting in undulating paths for the various waves. The waves, referred to as mode of propagation, will traverse the pipe in different times because the amount of refraction will vary. The time differences are not noticeable over a short distance, but they can cause distortion over a long distance. This method is called *multimode*.

Fig. 4.5. Fiber Optics Illustrated

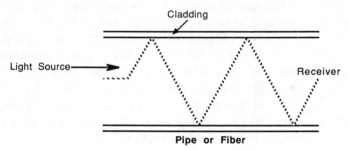

Now consider a lens that can focus the light beam into a very narrow band. If we also reduce the diameter of the pipe to complement this narrow beam, the light coupled into the pipe can be transmitted with almost no reflection. If the pipe is the width of a strand of hair, it is called a single-mode or monomode fiber.

Because there is little or no reflection, almost all the light coupled into the pipe travels along its optical axis. In other words, the transmit times for both the axially propagated light and the reflected light are approximately the same. This mode of transmission has always been preferred for passing information because greater distance can be achieved with less distortion than with multimode. The problem was the coupling of the units in the field. This problem recently has been solved, and single mode is now the dominant factor in fiber optics.

The major advantage of single-mode fiber is that it can operate in long lengths without repeaters (pulse regeneration) due to its low modal dispersion (that is, the lack of separation by the incoming optical signals). A single-mode fiber normally has a very small core diameter, which complicates the coupling of energy into and out of the fiber. However, high-speed lasers are now available to handle this requirement, and single-mode fibers have virtually replaced multimode. Another advantage of single-mode fiber is that the typical bandwidth is about 1000 MHz/km, whereas multimode is 10–20 MHz/km.

Light Transmission in Telecommunications

The use of light for transmitting information has been a dream of mankind probably since the start of messages. Within the United States, light transmission has been used from the inception of the country—"one [light] if by land, two if by sea." The introduction of the laser in the 1960s helped to stimulate growth in this area. However, only recently have we seen common applications of this technology as the field problems have been overcome and the need for more bandwidth has expanded. Consequently, the telephone network has been converting from analog to digital transmission, and digital pulses are employed easily in light transmission.

The analog wave is converted to digital and integrated with other digital messages up to the speed of the light source. The electrical pulses are converted to light pulses by turning the laser on and off. A simple example of this is a "1" represented by a red beam and a "0" represented by the absence of a red beam. The laser can be turned on and off thousands or millions of times a second to achieve the required rate. A photodetector at the receiving end will convert the light pulses back to electrical signals to be sent to the destination in the telephone network. A diagram is shown in Figure 4.6.

Several recent developments have greatly enhanced the fiber-optic business. One is the sudden move by the telephone industry from multimode to monomode (single-mode) fibers. With multimode fiber it was necessary to develop a way to compensate for undulating pulses along the core. This undulation would permit fast-moving pulses to overtake slower ones if the distance was sufficient. Repeaters were necessary at frequent intervals to prevent this problem.

Monomode fibers keep the pulses in a straight line, thereby reducing the chances for garbled information. This permits repeaters to be placed at much greater distances. New connectors have been introduced that have overcome the splicing problem of monomode fibers. Monomode fibers have thus become the standard for telecommunication.

Telecommunication has pioneered the introduction of optical techniques, due to the competitive nature of the business and because it realized that the days of copper wire, with its limited bandwidth, were numbered. The possibilities for the use of fibers within computers, communication systems, and eventually to the home are just beginning. The potential of this technology is enormous, and there is every reason to believe that homes will eventually be wired with fiber optics.

⌐ The main advantage of optical fiber over copper wire is bandwidth. Monomode fibers have virtually unlimited bandwidth, which means that one instrument on the desk will be capable of handling voice, data, video, facsimile, or any other communication needed by the user. The instrument itself will probably communicate internally via optics, thereby increasing the speed and capacity of the unit without increasing the size.

As previously mentioned, FDM uses filters to separate messages or conversations as they are received over a single transmission path. The same idea is being applied to fiber-optic transmission; filters are being

**Fig. 4.6.
Electrical to
Optoelectronic
Conversion**

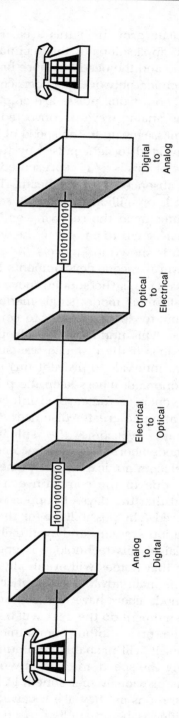

built that will allow two or more light beams to be transmitted at the same time over a path. The technique is called *wavelength-division multiplexing* (WDM). It will increase the already tremendous capacity of fiber-optic transmission.

Integrated Transmission Facilities

The basic light-wave technology currently operates at 90 Mb/s, a multiple of 45 Mb/s or DS3, and it can be used as the basic building block to explain light-wave signaling. The 45 Mb/s (actually it's 44.736 Mb/s) is capable of 672 voice channels, which is sufficient for most routes, except major national or international backbone routes.

The DS3 signals are normally derived from taking DS1 (1.544 Mb/s or 24 voice channels) signals and multiplexing four of them to a DS2 level (6.312 Mb/s or 96 channels). Seven DS2 channels are then multiplexed into a single DS3 signal. This method is the most common for deriving a DS3 in North America, although other increments can be used. This topic is explained further in Section 5.7 with TDM.

Not all of the bits are used for signaling, because synchronizing, parity, and control are needed to ensure that the information is being handled properly. However, any one of these schemes is possible, or combinations of the approaches can be incorporated into any configuration for deriving 45-Mb/s speed.

Many research and development (R & D) organizations are currently developing photonic (light) switches that will replace the current first-generation digital central office switches and will offer even greater capacity within the telecommunication industry. The future belongs to the optical industry, and we can expect changes to range from system architecture to the way we receive information within our homes.

4.4 Transmission Requirements

The transmission system is the distribution system or pipeline through which information is passed in a telecommunication network. Most transmission systems have little interest in the content of the information. Their main interest is the circulation of this information. Four key elements of transmission systems should be understood to fully appreciate how information is circulated: signaling protocols, transmission interface, synchronization, and system interface.

Signaling Protocols

Signaling protocols are what the transmission system uses to ensure that both ends of the connection are prepared to receive/transmit information. Signaling contains the various elements to prepare information to be transmitted. Signaling can be broken down into two general categories: customer line signaling and interoffice signaling. Line signaling is the interaction between the customer and the office or system serving the customer. Most people are familiar with these signals: they consist of ringing, dial tone, ring back, busy, answer, and other actions that might be involved in alerting or billing functions. Interoffice signaling is a more esoteric art, known only to a few members of telecommunications and makers of "black boxes."

Signaling protocols are being affected by deregulation, especially in the area of loop limits (the distance between the customer's location and the switching office). The "cheap" phones can have a problem on longer loops recognizing a signal to start ringing, or the switching office could have problems recognizing a request for service or the dialed or keyed digits. These signaling protocols are well known but cannot be taken for granted.

Telecommunication is introducing more computerese into its terminology. What is presently a simple call will become a session with the phone terminal. The customer initiates a session (lifts the handset); the switch indicates a session is permitted (returns dial tone); the customer addresses the switch (dials digits); the switch finds and alerts the requested party (ringing); or the switch supervises the termination of the session (everyone is disconnected).

One example of the difference between the customer's protocols and the interoffice signaling protocols is the transmission of digits in the form of frequencies. Table 4.1 compares customer tones with interoffice tones. The switching system is responsible for interpreting these frequencies and converting from one to the other where necessary.

Interoffice Signaling

For years the telecommunication industry has used an interoffice signaling scheme superimposed on the voice or link circuit. Many problems have been associated with this method of signaling, not the least of which is the black box gang or Captain Crunch's ability to beat the system and receive free long-distance calls. (The name "Captain Crunch" is derived

Table 4.1.
Customer and
Link
Frequencies

Digit	DIGIT CODE Customer Frequencies	Interoffice Frequencies
1	697 + 1209	700 + 900
2	697 + 1336	700 + 1100
3	697 + 1477	900 + 1100
4	770 + 1209	700 + 1300
5	770 + 1336	900 + 1300
6	770 + 1477	1100 + 1300
7	852 + 1209	700 + 1500
8	852 + 1336	900 + 1500
9	852 + 1477	1100 + 1500
0	941 + 1336	1300 + 1500

from the cereal in which someone discovered that the whistle in the box sounded like the disconnect tone on the long-distance network.) Another problem is the time lag from completion of dialing to connection. To be fair, this time lag was never considered excessive on long-distance voice connection. Alas, times change, technology moves forward, and today great criticism is heaped on post-dialing delays.

What will be missed from interoffice signaling is the terminology. One would think circuits or links between offices were busy or idle; not so, the industry requires a group of esoteric terms, the definitions of which were forgotten shortly after their creation. For example, idle normally meant off-hook, and busy normally meant on-hook, a carryover from customer signaling. Other terms such as hi-low, wink, and E & M will soon disappear as they are replaced by CCIS protocols.

A common-channel signaling method (CCIS or CCITT#7) will free the voice channel for voice/data calls and will speed up the completion of the calls. A comparison of the common-channel method with link signaling is shown in Figure 4.7.

The common channel or CCIS (common controlled interoffice signaling) will provide increased speed, greater capacity, better routing, and less coupling between signaling and conversation. In other words, a separate network for signaling between offices has been created. This network can grow independently of the voice/data/video network and its requirements. International groups are working toward this goal and they are to be commended for their fine effort.

A CCIS network originally was planned to be two-tier. The first tier would contain direct links, and the second tier would be referred to an STP (signal transfer point) and serve the same function as a tandem switch in the telecommunication hierarchy. North America was to be

Fig. 4.7. Link Switching Comparison with CCIS

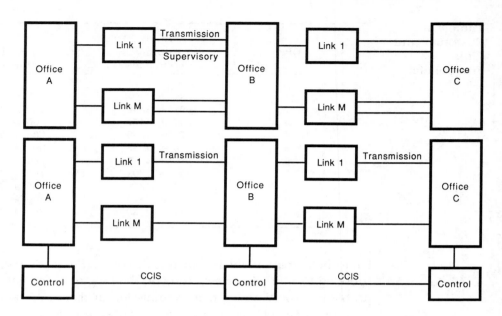

divided into 10 regions, but this plan was temporarily suspended when the Bell System was divided into its many parts. AT&T and the operating companies are now working to reintroduce a CCIS plan.

CCIS offers many advantages over current link signaling, but it costs more to implement. Additional data link facilities are required together with a connection to an STP for optimal routing. The expense is, however, worth it because this method of signaling will be very beneficial for future services. CCIS can provide access to data bases without going through the network. This would allow, for example, a check on 800 code status and routing.

Transmission Interface

The ability of the various transmission media—radio, fiber optics, wire— to interconnect is the result of a loss-design plan for the transported information (voice). The plan is implemented differently, depending on whether it is an analog or digital environment. To no one's surprise, I will discuss only the digital environment, which is a shame because the analog structure is a monument to man's ingenuity especially since it was constructed before the electronic age.

The loss-design plan is not the philosophy of the Chicago Cubs but a method for deciding the level of the voice between both ends of the connection. The loss is measured in decibels, the unit for sound. One dB is approximately the lowest level of sound, and it moves up from there, as illustrated in Figure 4.8.

Fig. 4.8. Decibel Range

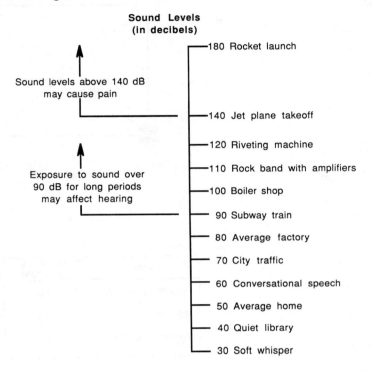

Sound Levels
(in decibels)

180 Rocket launch

Sound levels above 140 dB may cause pain

140 Jet plane takeoff

120 Riveting machine

110 Rock band with amplifiers

100 Boiler shop

Exposure to sound over 90 dB for long periods may affect hearing

90 Subway train

80 Average factory

70 City traffic

60 Conversational speech

50 Average home

40 Quiet library

30 Soft whisper

As we speak, the dBs we transmit will encounter some loss before they are received. A phone conversation tries to emulate two people conversing in a room and standing about 20 feet apart. The loss within the room will vary depending on the conditions, but the loss should range from 4 to 10 dB. Most people expect a higher dB loss with distance, although with a digital network it can be avoided. The plan is known as loss-design because it controls the dB loss in the connection from end to end.

For a digital environment a fixed-loss plan was proposed by AT&T. Under this plan the end or Class 5 offices operate at 0 dB. The digital Class 4 or higher offices operate at − 3 dB, and an intertoll connection would have a fixed loss of 6 dB. Figure 4.9 shows a two-tier switch with the related loss for each link in the connection. Figure 4.9 shows both the recommended and maximum losses for various offices within a digital network.

For a digital environment the loss can easily be controlled, whereas in the analog environment substantial work was needed to maintain the proper transmission levels. Where there is a mixture of digital and analog, the transmission levels must be known at each conversion point. A

**Fig. 4.9.
Transmission
Loss Plan**

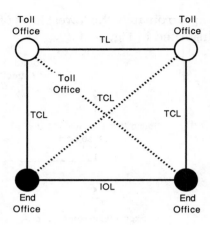

Link	Recommended Loss (dB)	Maximum Loss (dB)
Interoffice Link (IOL)	3.0	5.0
Toll Link (TCL)	3.0	4.0
Tandem Link (TL)	0	

couple of terms have been introduced to assist in this determination: Encode level point (ELP) refers to the level of the analog signal before it is encoded to digital; decode level point (DLP) refers to the level of the digital signal after it is decoded to digital. These levels are important in meeting the transmission objectives where combined analog and digital networks are in place.

Synchronization

A digital network, similar to other networks, requires systemwide synchronization to preserve the bit integrity of the transported information. It provides bit sampling at the same rate throughout each office within the network. The logical place to synchronize is at the least common scan rate for all offices, presently 8000 Hz. The basic signaling rate within a digital office is DS1 or 1.544 Mb/s. The bit rate within a DS1 system consists of 24 voice channels sampled 8000 times per second with 8 bits per sample plus 1 framing bit 8000 times per second: thus $24 \times 8000 \times 8 + 1 \times 8000 = 1,544,000$ b/s $= 1.54$ Mb/s.

The Bell System has what they call a Bell System reference frequency standard, which is a master timing supply from an atomic clock located in Hillsboro, Missouri. This supply keeps all offices synchronized to a common 8000-Hz supply. Consequently, all switching within all digital offices is taking place at a DS1 level driven from a single source. This is shown in Figure 4.10.

**Fig. 4.10.
Synchronization**

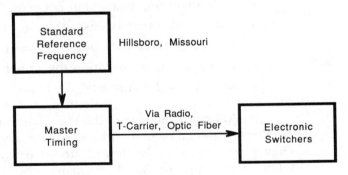

Switching system timing in the long-distance network can be traced to Bell's reference frequency. The importance of this is that anyone wanting to establish an independent network must decide whether to drive the network from their own source or use Bell's source. It is easier to drive the network from Bell's source, but a T-carrier interface is normally required to establish the source. In this case an agreement with Bell is needed. Independent networks, even if sourced from Bell's reference, are equipped with their own internal clock. Hence if the Bell reference frequency is lost, the network can continue to operate for several weeks from its own source.

System Interface

The interface between the switching system and the transmission equipment was well defined during the analog period of telecommunication. The switching system was responsible for everything on one side of the mainframe, and the transmission equipment for everything on the other side. Interconnecting switching and transmission was a unit called a *trunk* or *link*. The common term for years was trunk circuit, which meant the communication or transmission channel between two separate switching systems. The trunk facility referred to the transmission media between trunk circuits. This term has virtually disappeared from telecommunication, and *transmission media* has become the accepted term.

The trunk circuit was really the means whereby various types of equipment could interface. In modern terminology a trunk circuit might be called a protocol converter for interfacing switching systems from different vendors.

The trunk circuit

- Controlled the connection between two offices
- Furnished transmission battery to the originating party
- Repeated the called number toward the destination
- Repeated the information the called party has answered
- Initiated and terminated the billing process

The various functions of the trunk circuit have been moved to other portions of the switching and transmission equipment. The start of this change was the introduction of stored program control (SPC) systems where the computer could perform many of these functions, thereby saving equipment in each trunk. The introduction of common-channel signaling is removing the last function from the trunk circuit, and the trunk, for all practical purposes, will vanish as the network becomes digital. Hence, the term "trunk" seems inappropriate to describe the connection between offices. The term "link" appears to be proper for current telecommunication and is used throughout the descriptions.

The introduction of digital switching and digital transmission over the years has done more to merge these two functions than any analog effort. A brief example of how this demarcation line has blurred recently might be helpful in understanding why switching vendors and transmission vendors are seeking the same market.

The arrangement for the connection between the switch and the facility when the digital carrier was originally introduced is shown in Figure 4.11A. As illustrated, the separation between switching and transmission still was the trunk, and the trunk was part of switching. The transmission vendors realized that they could provide their customers with some economies and gain a new market for themselves by offering units that would combine the trunk and the channel banks as shown in Figure 4.11B.

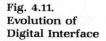

Fig. 4.11.
Evolution of
Digital Interface

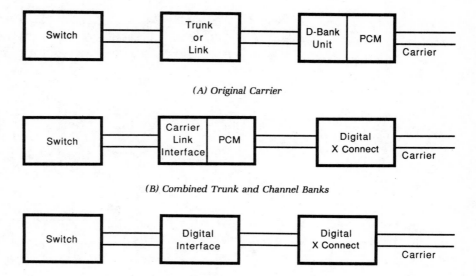

(A) Original Carrier

(B) Combined Trunk and Channel Banks

(C) Digital Switch Interface to Carrier

The introduction of the digital tandem office and the digital PBX changes this approach, because the office is now switching calls at a digital rate and the interface to the carrier now moves to the switching side. The typical arrangement for the digital switch interface to the carrier is as in Figure 4.11C.

The digital switch has combined the switch interface function with the channel bank functions, thereby simplifying the crossconnect scheme and reducing the cost. These connections are the predecessors of the digital network of the future when many of these connections will disappear as the interface between switching and transmission blurs.

The arrangement in Figure 4.11C has introduced a new area for switching within telecommunication—digital crossconnect. The digital crossconnect system is a switching device that interconnects the T1 digital subchannels—64 kb/s DS0 lines. It is normally controlled by software and accessed from a standard computer terminal. It can rapidly crossconnect any pair of T1 subchannels, thereby permitting different functions with the T1 lines.

5

Data

We enter the lair of the ogre called data where many attempts, with few successes, have been made to define and harness this beast. Obviously, much has been written on the subject, and any attempt to offer a clear understanding within a few chapters is naive and foolish. However, we must be aware of some basic concepts to build a network.

5.1 Introduction

At this time in our history the image of data is stamped on everyone's mind. In the 1960s data was not clearly understood: people still used datum as a singular form of the word. Predictions were made that data would account for 50% of the traffic on the telephone network by 1975. We abandoned our concepts that data was information in table or chart form and searched for this intriguing new phenomena. We were deceived.

We spent the 1960s and 1970s breaking down data into smaller and smaller tasks and made it duller and duller, attempting to make it subservient to voice transmission. We told the computer market that a transmission rate of 300 c/s was good enough for most data. Not until we started to break voice down to a "pure" information state did we realize that the notion of data for data's sake was very important to our industry.

The telephone industry now realizes the value of expressing both voice and data in an information form. It would really like to be known as the information industry and have the term "telecommunication" understood as voice, data, facsimile, and video. Why the change? I think part of the reason is a deeper insight into the world of communication brought about by the information age. A better answer might come from paraphrasing the old philosophical saying: "A tree is a tree; a tree is not a tree; a tree is a tree" into "data is data; data is voice; voice is data; voice and data are information."

5.2 Data Information

Serial and Parallel

The basic building blocks of data are binary codes of pulses represented by 1's and 0's. There are several ways to represent these pulses—on/off units, positive and negative voltages—but the bottom line is a 0 or 1, the basic elements of a binary code.

The transmission of this information can be in serial or in parallel modes. The serial method transmits 1 bit at a time. Figure 5.1 shows a 7-bit word (1010110) being transmitted serially.

Fig. 5.1. Serial Transmission of Information

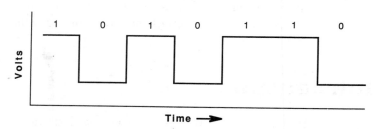

The serial method is usually employed in telecommunication between offices within the network. The primary user of the parallel mode is the computer. Originally all transmission between a computer and its peripheral equipment was in a parallel, binary mode. The parallel mode for the number 1010110 is shown in Figure 5.2.

All 7 bits are transmitted simultaneously. The introduction of data networks and the need for computers to transmit over great distances have reduced the amount of parallel transmission. Most networks transmit information serially, although there are other ways.

Control Procedures

Control procedures are needed for operating a communication system. Basically, these procedures are called codes and protocols within the computer industry. For telephone systems the code system was the voice, and

**Fig. 5.2. Parallel
Transmission of
Information**

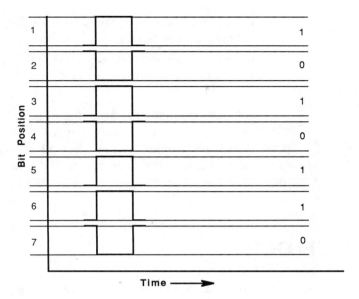

the protocol was the signaling between offices or between office and subscriber. This protocol was established before deregulation and controlled by
the Bell System for North America. In addition, telephone systems use dial
pulses and MF (multifrequency) tones for routing control.

The codes in computer communication are the 5-, 7-, or 8-bit characters and numerals received by the terminals or mainframe. The protocols are sets of rules for controlling the system. During the 1960s the
codes and protocols were part of the same information. For example, the
American Standard Code for Information Interchange (ASCII) has 128
valid arrangements, including the characters and numerals transmitted.
Also included in this 128 count are control procedures such as "start
text," "end text," and "cancel." More recently, the computer industry separated the codes from the protocols, reflecting the growth of the business.

Protocols have been established by manufacturers and standard
committees within the industry. For our purposes the protocols are
designed to solve framing problems, provide error and sequence control,
and interpret special timings. The current protocol is bit oriented, which,
as previously stated, defines those bits that are the message. It is accomplished by flags inserted around the message to delineate it. Several bit-
oriented protocols are available; some are

- IBM synchronous data link control (SDLC)
- American National Standards Institute advanced data
  communication control procedure (ADCCP)

- International Standards Institute high-level data-link control procedures (HDLC)
- International Telephone and Telegraph Consultative Committee (CCITT) recommendation X.25

The first bit-oriented procedure, SDLC, was introduced by IBM in 1969, and the standard groups were close behind. In general, these procedures are similar in that they place barriers or flags around the message and allow the message to have variable length. This does not mean that they are compatible with each other.

5.3 DTE/DCE

Access to and control of information are extremely important. The business world is inundated with terminals connected to something. The most visible part of a terminal is the input/output portion. The standard input today is a keyboard, which looks like a normal typewriter and is basically used the same way. Each keystroke is captured as a digitally coded signal used by the terminal for transmittal to the data processing host.

Output to the terminal or user is accomplished in several ways: hard copy from a printer, soft copy displayed on the screen, or both, depending on the requirements.

The terminal or DTE (data terminal equipment) can use the 3-kHz telephone channel for rates of 1200, 2400, 4800, or 9600 b/s. The speed is determined by the DCE (data communication equipment). This unit is normally separated from the DTE to give more flexible arrangements. The interface is defined in Electronic Industries Association (EIA) Standard RS-232 or CCITT Recommendation v.24 for the 25-pin plug and socket. For 9600 b/s the line must be conditioned—that is, it must have no bridge taps or loading coils—and have private access.

The DTE is normally the equipment associated with the user (typewriter, terminal) or being communicated with (mainframe) and operating in a digital format.

The function of the DCE is, for modems, to convert the digital input to analog for operation on the network. In a data network operation the DTE functions to interface the various devices.

The requirements for DTE and DCE within a company primarily depend on the use of data. There is no doubt that a company using data only to handle invoices locally will require different DTEs and DCEs than the company that is decentralized and uses large data processing systems for development. For the former a simple transactional system tied directly to the computer or through a concentrator will satisfy the needs. However, for the latter an information system with an integrated data base and various DCEs—possibly a packet network—is required.

The information system problem is broadened when a company's locations become dispersed. In this case a network architecture such as IBM SNA (systems network architecture) is required, and attention must be paid to the access methods. The relationship between the DTE and the DCE can become blurred; in many applications both are combined into one unit. Many data processing situations do not want to take advantage of these combined facilities for their network. Many applications want the interface separate from the terminal to cater to technical improvements or economic advantages in either the DTE or DCE.

5.4 Transmission of Data

Information Coding

The most common code for transmitting information is a binary code or binary pulses. For *n* pulses there are 2 different combinations available in a binary system. For 5 pulses, 32 different combinations can be derived; for 6 pulses, 64 are available; for 7 bits, a common method, 128 combinations can be transmitted.

A common method of error checking was the addition of a parity bit added to the character stream to indicate, for example, whether the number of 1's within a word is odd or even. This scheme is still available, as is the bit-oriented checking scheme, such as SDLC.

Bit and Baud

The transmission media can be arranged to transmit information in analog form or digital form. With digital form the information content consists of combining the baud rate and the bit rate to determine exactly how much information can be sent per unit of time.

Baud rate is frequently confused with bit rate, often by people who should know better. The *bit rate* is the basic measure of the information or message being transmitted. For digital it is the 1's and 0's of the information. The *baud rate* is the measure of the signaling speed. When the information rate is 1 bit per baud, the two are equal, but there are many schemes that allow the bits per baud to be greater than 1. These situations cause confusion because now the baud rate is, say 1200, whereas the bit rate can be 2400 or 4800, with 2 or 4 bits per baud being transmitted. For voice transmission with pulse code modulation (PCM), the bit and baud rates are normally the same. However, this has not been true for data, because there are schemes to increase the bit rate without changing the baud rate. Consider the arrangement in Figure 5.3.

Fig. 5.3. Bit and Baud Comparison

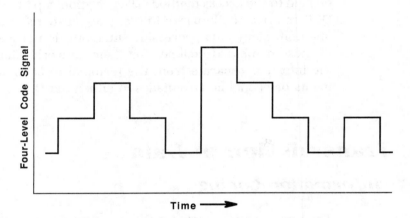

The information rate is equal to the information per signaling element (bits/S.E.) times the signaling rate or the number of signaling elements per second (baud). With two states (0 or 1) the information per signaling element is 1; with four states, as shown in the figure, the information per signaling element is 2 (the bits per signaling follows log L, where L is the level being transmitted). The information rate in this case will be twice the rate when there are only two states, assuming the baud rate is the same.

Synchronous versus Asynchronous

Synchronization is the method whereby a precise time relationship is maintained between the sending and receiving terminals. The two basic forms of operations in this area are *synchronous* and *asynchronous*.

Asynchronous transmission is characterized by the creation of start (or setup) bits and stop bits. It is often referred to as start/stop transmission and used with a code such as ASCII. This permits the interfacing of data equipment operating at different speeds and allows machines with no buffer to interface. The arrangement is shown in Figure 5.4.

Fig. 5.4.
Comparison of
Asynchronous
and
Synchronous
Transmission

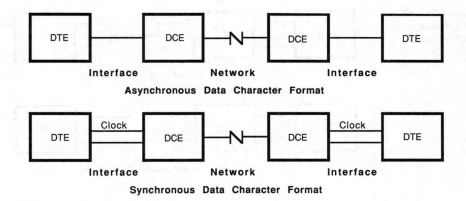

Note the clock lead in addition to the information lead between the DTE and the DCE for synchronous transmission. This lead is not required with asynchronous operation. It allows synchronous operation to eliminate overhead and, theoretically, provide faster transmission. However, it can add complexity to the equipment and restrict the interface to other equipment. Asynchronous can provide interface to more terminals. Synchronous transmission operates with the protocol schemes previously mentioned and is very useful for building data networks.

Half-Duplex and Full-Duplex Transmission

Terminal equipment can be designed to transmit in both directions at once (full-duplex) or in one direction at a time (half-duplex). Which method is employed often depends on the transmission line between units.

Half-duplex operation permits communication in only one direction at a time between two stations. Schemes have been devised to overcome this shortfall and permit simultaneous operation. The most common scheme is the use of separate frequency bands in each direction. Proper frequency selection prevents the frequencies from interfering with each other.

Full-duplex allows both ends of the connection to simultaneously receive and transmit. It normally requires a four-wire circuit. The advantage of this method is its ability to transmit data streams in both directions, or data in one direction and control signals in the other. The control signals govern the data flow and perform error checking.

A comparison of half-duplex and full-duplex is shown in Figure 5.5.

Fig. 5.5. Half-
Duplex and
Full-Duplex

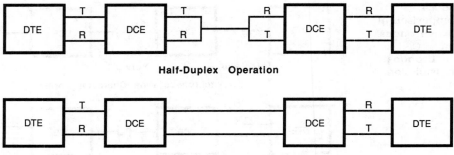

Half-Duplex   Operation

Full-Duplex   Operation

5.5 Modems

Many data transmission operations have only occasional need to transmit
information. For these occasions it is appropriate to connect data communicating devices for transmitting low- or medium-speed data across the
voice-grade dial network. This network is in place and providing service
over a nationwide system of switched lines, and the only interface device
required is a modem at both ends. The modem presently is the universal
device that meets the reliability criteria of the network and the service
requirements of the customer. The call is charged at the same rate as a
voice call. Although it is reasonable for infrequent calls, this characteristic
has created the push for special data networks among large users.

To transmit digital signals over the analog channels of the voice
network, which passes frequencies of 300–3000 Hz, a transmitter is
needed to modulate a voice-frequency carrier signal from the data, and
a receiver is needed to demodulate this signal. Consequently, this data
transceiver is known as a modem. The principal application of modems
is to interconnect data terminals with communicating channels. A
modem also performs control functions to coordinate the flow of data
between the terminal equipment. These control functions can be compared to the supervision of a telephone call: ringing, answer, disconnect, to name a few. Many of the checks for good continuous
transmission in a phone connection are performed by the parties engaged in conversation. In a data call between modems, these checks
must be done by the units.

The early modems used the interrupting of the 20-milliampere (mA)
signal on the line to transmit information. The most common modem
today uses a transmitting standard derived from EIA RS-232, which is
based on different frequencies for information.

Typically, modems operate at 300 b/s or baud (and higher) and use frequency-division multiplexing (FDM). The telephone allows frequencies from 300 to 3000 Hz, but the modem avoids the 2500–3000 Hz, because 2600 Hz is a disconnect tone for the network. Hence, tones are selected from 300 to 2400 Hz. To achieve full-duplex operation, modems are designed with a set of originating mode frequencies and a set of terminating mode frequencies. Variations are available, but we will stick with the norm. The Bell System 108 modem, for example, transmits a tone of 1070 Hz for a Space and 1270 Hz for a Mark. It interprets the reception of a 2025-Hz tone as a Space and a 2225-Hz tone as a Mark. The distant modem will interpret the 1070-Hz tone as a Space and the 1270-Hz tone as a Mark. It will send a 2025-Hz tone as a Space and a 2225-Hz tone as a Mark.

The speed of modems became a critical factor by the late 1970s, and the industry introduced phase modulation to synchronous modems, allowing a 1200-baud operation and a 2400 b/s information rate through the use of *dibits*. Dibits (00, 01, 10, 11) are transmitted during phases (e.g., 45°, 135°, 225°, and 315°) of the transmitted sine wave. The introduction of flexible software to adjust to various standards within the industry has also pushed the use of modems into new applications.

Today, two modem specifications dominate the field: CCITT's and Bell's. The Bell specification, known as EIA RS-232, is used in North America and has allowed other modem manufacturers to enter a field once completely dominated by Bell. The EIA has been working to enhance this specification as the range of applications for modems increases. For example, RS-232 does not work well over long distances, so RS-422 was introduced. For sites with different voltages RS-423 has been approved. High-speed data connections should follow the RS-449 specification. These enhancements, together with large-scale-integration (LSI) chips to handle the complex sections of the various interfaces, have helped to make modems widely used.

The v.24 CCITT specification is used throughout Europe. It basically differs from the Bell specification in line levels and telephone energy bandwidths.

Two basic types of modems are available: dial-up and dedicated. Dial-up modems operate over the standard telecommunication network, typically at speeds of 300 and 1200 baud. Dedicated modems employ unique twisted pair or coaxial cables and operate at speeds of 19,200 and 38,400 baud. The dial-up modem is starting to dominate the marketplace as advancing technology lowers the price while increasing the speed. The latter is offering almost direct competition with the dedicated modem. As the modem is speeded up, new applications are emerging for their use.

The modem has become the first commodity item of the new voice/data telecommunication.

Choosing a modem can be a complex process. Modems will provide half-duplex or full-duplex operation. It probably is smart to select the full-duplex operation even if the half-duplex would suffice; the cost differential is not that great, and the full-duplex provides more flexibility for future options.

An asynchronous operation is probably sufficient, because it provides interfacing to more devices and the overhead is not noticeable at low speeds. Asynchronous operation has been greatly helped by the personal computer (PC) market as manufacturers attempt to supply this ever-growing field.

5.6 Multiplexing Techniques

A popular method for bridging the difference between terminal speed and transmission media speed is multiplexing. That is, the signals are sent simultaneously from many terminals over the same transmission line. Multiplexing is of great economic value because the operations that are multiplexed are so slow compared to the operating speed of the media. Two multiplexing methods available to the average user are FDM and TDM.

Programmable formats and speeds enable a multiplexer to answer a dial network call and adjust to the characteristics from the remote end of the call. The interface requirements for multiplexers are equivalent to the standards of the modem, thereby allowing the design of these units to focus on compression and statistical techniques.

Frequency-Division Multiplex

Time-division multiplex was the original method of deriving additional communication channels from a transmission media but, due to practical reasons, didn't achieve success until the introduction of integrated circuits (ICs). Frequency-division multiplexing for analog systems was the modulation carrier of the 1960s, because the implementation techniques were easier and further developed. Frequency-division multiplex assigned different conversations (each with 4000 Hz) to different segments of a high-frequency band, thereby creating bands of frequencies. By using each frequency band as a separate channel for transmitting information, we could transmit many calls over a single media. Electrical filters are used to position the bands for transmitting and to separate the bands at the receiving end. The calls are all transmitted at the same time, but they occupy different segments of the frequency.

Time-Division Multiplex

As more and more information is coded into a data format, the need to transmit billions of bits of information almost instantaneously increases. The number of channels available for this transmission is limited, and some type of allocation system must be used. Modern technology is introducing fiber-optic links capable of transmitting billions and billions of bits of information (a billion bits is known as a gigabit) over great distances. Independent channels making different demands on high-speed channels (e.g., fiber optics) or on limited cable pairs in a congested area leads to an allocation scheme known as TDM.

When interconnecting signals at different transmission rates, a multiplexer performs different functions, including signal formatting and synchronization. Signal formatting is the coding/decoding necessary to interface between the method used in the terminal and the method used in the multiplexer.

Synchronization is necessary because the low-speed signals received from terminals do not quite match the high-speed rate of the multiplexer. As the use of multiplexers increases, the ability to interface with various types of terminals with different speeds will become more necessary.

Time-division multiplexing requires that each call occupy the transmission media for a small segment of time and then allow the next call to occupy the media for its segment. The distance end is synchronized so that the segments are removed at the proper time. Both ends of the connection scan the input/output in a rotary-motion fashion to coordinate the passing of the signal between the proper time slots.

Network arrangements for multiplexers are covered in the next chapter.

5.7 Pulse Code Modulation

The voice, the principal item transmitted across the telecommunication network, is transported in the form of a sinusoidal wave whose signal is continuously attenuated (energy loss) as it progresses toward its destination. At periodic intervals the signal is amplified to compensate for this attenuation. However, any noise that has been accumulated along the way is also amplified, causing the signal to be distorted: the longer the connection, the greater the distortion. In the extreme case the received signal could be unintelligible because the amplifiers cannot distinguish between noise and actual signal.

The attraction of digital methods for dealing with voice transmission is their potential for eliminating the distortion caused by amplification. The advantage of digital transmission is that the signal exists only in two states—0 and 1. No matter how distorted the signal is, as long as there is the presence or absence of a state, the system can handle it. As it progresses along the transmission medium, the signal does not require amplifiers but repeaters that can detect a pulse at the input.

The obvious advantage of transmitting data in this format is that data originally was in this format and no conversion to analog is necessary.

The process of turning the analog signal into a stream of pulses is known as digital encoding. If properly performed, it will allow the original analog signal to be recovered without significant impairment. In telecommunications this process is called PCM.

The digital encoding process is divided into sampling and quantization. *Sampling* involves taking microscopically brief samples of the analog signal at exactly regular intervals. Pulse code modulation uses a sampling rate that is double the hertz rate of the analog signal. Although an analog signal for voice is approximately 3400 Hz, for practical reasons it is assumed to be 4000 Hz, and a sampling rate of 8000 is used.

Quantization involves the measurement of the size (voltage) of each of the successive signal samples of the voice or sine wave. This value is measured and expressed as a binary number. The technique is shown in Figure 5.6.

For voice sampling, a code involving eight binary digits is sufficient for accurate reproduction. The present PCM technique, therefore, requires 8000 samples per second (samples/s) and an 8-bit sample to transmit the representation of the analog signal, yielding a 64,000-b/s rate (8000 samples/s × 8 bits/sample = 64,000 b/s or 64 kb/s).

Other encoding techniques are available, but PCM is the recognized standard for telecommunication. The present network is a mixture of analog and digital transmission, and no other method can encounter the number of transformations from analog to digital, and vice versa, as well as PCM.

Typically, central offices are analog, and the facilities between these offices are digital, which necessitates the constant conversion from analog to digital and from digital to analog. Once the network is converted to

Fig. 5.6. Pulse Code Modulation

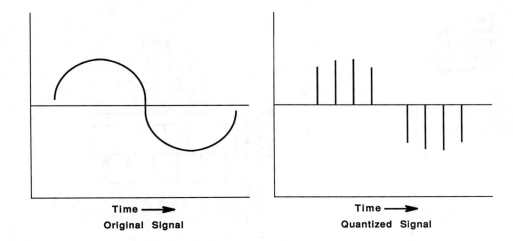

Time ⟶
Original Signal

Time ⟶
Quantized Signal

digital (or nearly so), the sampling technique will probably be changed. More economical approaches to encoding are available; for example, there are excellent methods of compressing voice to 32 kb/s or 16 kb/s, but, because of the mixed network, 64 kb/s will be the standard for a while.

5.8 Hierarchy

Frequency-Division Hierarchy

The FDM hierarchy consists of four levels, with the bandwidth for each level successively increased to permit additional bands for calls. The levels are shown in Figure 5.7.

Time-Division Hierarchy

Bell introduced the T1 carrier in 1962 and began PCM electronic switching. The T1 system allows 24 voice channels to be multiplexed onto a single transmission media. The benefits of PCM were seen immediately, especially in large cities where cable growth was restricted. In those applications the cost of the terminal equipment was low compared to the gain in voice connections. A simplified diagram of the T1 system with binary pulses is shown in Figure 5.8.

Each voice connection during its time on the transmission media transmits eight binary pulses; therefore for 24-voice connections 192 (8

Fig. 5.7.
Frequency-
Division
Multiplexing
Hierarchy

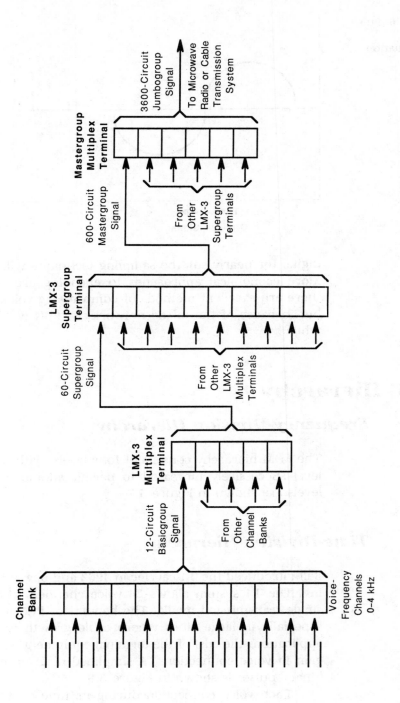

Fig. 5.8. T1 System

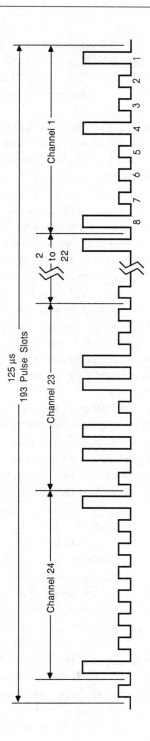

× 24) binary pulses are transmitted. There is also a 193d bit used for framing or synchronizing both ends of the carrier. These sets of pulses are transmitted 8000 times per second for a total of 1,544,000 (193 × 8000) pulses or bits per second (1.544 Mb/s). This is also known as the DS1 rate.

The T2 system was introduced by Bell in 1972 to carry 96 voice conversations on a single cable pair or on one Picturephone connection. Each level of the digital hierarchy relies on an increase in speed. A 6.3-Mb/s transmission rate (DS2 rate) was used in T2, but at the time it was introduced it was not considered quality video. However, it was sufficient for the voice connections over a wire pair, and the terminal equipment was relatively inexpensive compared to the voice connection gain. The T1C level was also introduced and operated at 3.152 Mb/s (DS1C rate), which allowed 48 voice conversations to be transmitted over the cable pair. T1C was probably introduced because early T1 systems were engineered conservatively, and at least twice the distance could be achieved without degrading the quality of the voice connection. Another answer was to double the conversations over the same wire pair without changing the regenerative repeaters. T1C has recently become extremely popular in applications where T1 is used, but an increase in traffic is expected. The incremental cost increase for T1C is small, and it can act as a step toward T2.

The next step in the digital hierarchy is the T3 or DS3 level, which can carry 672 voice connections and is designated as one digital mastergroup. A transmission rate of 44.736 Mb/s (DS3 rate) is used. The DS3 rate was found to be too fast for a wire pair and too slow for coaxial cable, so for years its principal use was as a bridge between T2 and T4. However, fiber optics has increased the use of the 44.736-Mb/s rate, because optical fiber cable systems have virtually unlimited bandwidth capacity. The optical fiber system employing the DS3 rate is normally designated as a 45-Mb/s or 90-Mb/s (two separate DS3 connection) system and will usually interface to DS1, DS1C, DS2, and DS3 rates.

The T4 carrier operates at 274.176 Mb/s (DS4 rate) and can transmit 4032 voice conversations. The T4 carrier generally uses coaxial cable for transmitting, although optical systems may change this approach. The present digital hierarchy is shown in Figure 5.9. A T5 rate is being investigated due to the capabilities of fiber-optic systems.

The ability of the digital format to transport information without encountering any noise problem is one of its unique advantages, and present light-wave systems are built on digital format. Interconnecting high-speed light-wave systems will introduce different levels within this hierarchy as more and more facilities are converted to optical systems.

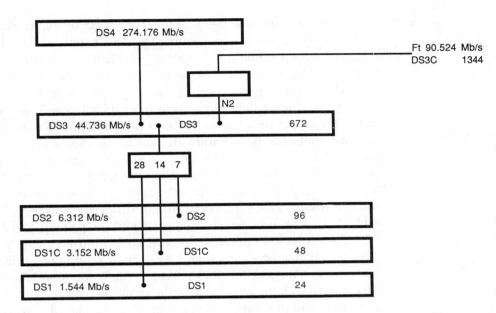

Fig. 5.9. Time-Division Hierarchy

To implement this approach within the network, we set the digital lines at the 64-kb/s rate and grouped them into the 1.54-Mb/s T1 rate. However, for data the terminals are operating at different rates and must be synchronized or nearly synchronized throughout the connection.

5.9 Voice/Data Integration

Before personal computers the main business devices for communicating were telephone and mail. The communication received via mail is changing rapidly to information via the computer with its own network or networks. The evolution of these sources of information is that a single network will emerge to serve both data and voice. The merging of these two functions into an integrated network requires a radical change in the facilities presently employed and in the switching systems. Most corporations have merged voice and data departments hoping that these two groups will be able to share facilities and eventually systems. This merger is an effort to solve the problem of ever-increasing communication costs.

The eventual combining of voice and data will be via digital switches and digital facilities; however, the present world is considered to be analog switches and analog facilities, although digital switches are starting to predominate. As rates become more attractive for digital facilities, they will lead to an accelerated schedule of replacement.

There are other aspects of this voice/data problem before a solution is found. Voice switching requires a circuit-switched facility at 4 kHz, whereas low-speed data can operate effectively on a packet-switched network and high-speed data needs a substantial bandwidth beyond 4 kHz.

The present generation of PBXs (private branch exchanges) offers some type of voice/data integration but typically needs a special workstation with a data terminal driver and unique interfaces at the PBX. These arrangements permit a bandwidth up to 56 kb/s, more than enough for everyone but the most demanding user. This solution is expensive, however, and if the data users are a select group a data multiplexer for those terminals may be a more viable solution. The multiplexer will probably require the installation of a coaxial network for the user, but, thanks to cable TV, coax is relatively inexpensive, and the principal cost will be labor.

The expected approach for the integrated PBX will be to offer one switch that has processors and associated equipment imbedded within the system. These processors will most likely independently handle voice (circuit switching), low-speed data (packet switching), and LAN (local area network) interface (Ethernet or ring compatible). In addition to the problems associated with the handling of these three techniques, the PBX will be called upon to employ the most comprehensive management system, because the PBX is now a sine qua non part of the business. If PBXs reach the level of importance described here, the tolerance for error must be extremely low both locally and end to end.

Other features that this world PBX must offer include least-cost routing or equivalent and interfacing with mainframe computers. Incorporating these features into a simple voice PBX appears to be a herculean task, but developments are taking place that will eventually do this.

5.10 Special Digital Services

The use of digital facilities throughout the country allows digital services to be used for special transmission requirements. These services consist of T1 switched service, T1 full-period service, packet service, or dataphone service.

The T1 switched service is offered, for example, by AT&T as an end-to-end high-capacity 1.54-Mb/s digital utility that will provide videoconferencing or other special voice/data communication. The facility is carried from the customer's premise to an office, where it is switched over terrestrial or satellite links to the terminating location. This is AT&T's

ACCUNET service, the first of many T1 services we will see in the next few years.

As bandwidth capacity is added to the network through satellites or, more likely, fiber optics, T1 switched service will become routine. It is also expected that the price will decrease as facilities become available. Although T1 teleconferencing service will economically prove-in against the cost of travel, the major use of it will be to improve the productivity of knowledge sources within a business by making them available, live or via tape, to a greater audience than is presently possible.

If there is insufficient data or video transmission to justify a shared or full-time T1 line, the customer has the option of packet service. Packet service consists of bundling small units of data into a packet or envelope for transmission. The advantage of this is that a packet can only occupy the network during the actual transmission of the data, whereas with circuit (voice) switching the facility is occupied for the entire conversation regardless of whether conversation is taking place.

Packet service is available in a variety of forms and speeds, depending on the requirements of the customer, although there are two basic capabilities: virtual call and permanent virtual circuit. *Virtual call* provides call setup on a per-call basis; *permanent virtual circuit* is a full-period connection where no setup is necessary. Different options exist within these capabilities, but they normally vary by vendor.

Services are now available to the data user that did not exist several years ago. The customer can elect data service from several vendors and obtain information on which service meets the data requirements of the company. It is worthwhile to take advantage of this market-driven service.

6

Data Configurations*

In this chapter data is regarded not as a compilation of techniques to be contemplated but as a transport system for the movement of information. It appears that the introduction of this "system of movement" was to make clear the nature of data and show how it differed from voice.

6.1 Introduction

The basic concept of the original data networks was to interface data terminals with other terminals or mainframes as easily as phones are interconnected on the voice network. Originally the mainframe only communicated with peripheral devices (printers, disks, typewriters, CRTs) in streams of characters (parallel format). The information could be transported only a short distance with any reliability. The first step toward improving this communication was to convert the information to a serial format and transport it across telephone lines or directly. This permitted the peripheral equipment to be placed at greater distances from the mainframe, but many people found the voice network to be slow and error prone for this type of information. The greatest drawback was that to transmit a small burst of data every few minutes the connection had to be continuously held. A fresh approach was needed.

Data networks became available based on the circuit-switching technique of voiceswitching. The DATRAN network is a circuit-switched data network. However, many users felt that a voice-switching system was unsuitable for data, and different switching methods were needed to cater to data communication's bursty nature.

In 1968 Paul Baran reported in detail on a fully distributed packet network that could be used for all military communication, both voice and data. The packet networks that have emerged thus far have only dealt with transporting data. (I hope you realize by now how critical the late 1940s and the late 1960s were to the present. Does this mean the late 1980s are critical?)

Packet-switching functions are distinct from what is normally used in telecommunication—circuit switching. Circuit switching is the establishment of a permanent connection for the duration of a conversation between two parties or similar calls. It is commonly called *full-period transmission* with circuit switching.

For many types of data calls the full-period circuit is not necessary because the transmission consists of "bursts" of data passed back and forth with long periods of silence between them. This is especially true between a computer terminal and the mainframe. It would be a waste of facilities to assign a full-period circuit between a terminal and a computer that only required a connection for a few seconds. For such connections packet switching was developed.

A *packet* is a method of formatting the data information into blocks of characters and transmitting them across the network. The packet includes address, device speed, and other pertinent information necessary for proper routing and interconnection. Routing instructions pertaining to the call are provided to each node toward the destination. As each of these nodes receives a packet associated with the call, the route to the destination is established. This procedure will be followed for the entire call unless congestion or failure occurs in the route. If failure or congestion does occur, the node is able to route the packet via alternate routes until the trouble is corrected or the call is completed. The present packet networks, with their ability to have packets arrive in sequence at the destination, have substantially less overhead than do the original networks with scrambled packets. Packets have also improved the throughput (call-carrying) capabilities of the switches.

A unique aspect of a packet network is its ability to use different links for transmitting different packets of the same message. It also means that the packets must be reassembled at the destination before the message is acted upon. This approach can be compared to balancing a checkbook. The checks are written (transmitted) in sequence, but they are received at the end of the month in random order and must be reassembled (using the check number as the packet I.D.) before the checkbook can be balanced.

A packet-switching diagram for transmitting messages is shown in Figure 6.1.

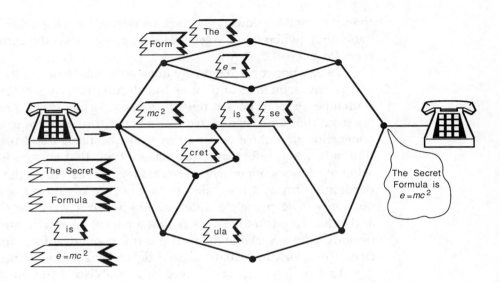

Fig. 6.1.
Example of a
Data Network
Operation

Packet switching can economically handle any type of switching presently known, including burst switching, store-and-forward switching, and voice switching. With regard to the latter, many studies are being conducted to determine methods for implementing voice communications on a packet network. The main problem is that certain packets can encounter longer delays and conversation could become garbled quickly. This problem does not appear to be severe when the voice messages are carried on a LAN (local area network). However, voice packets transmitted across the country still can encounter variable delays.

The advantage of using a packet network for all messages is the efficiency of the network itself. The network is used only when a message must be transmitted, which is a tremendous productivity improvement. Consider that during a typical conversation a customer uses the circuit only 40% of the time (50% if both parties divide the conversation equally); the other 60% is dead time. If that circuit was fully utilized by packet switching, we would achieve a 150% increase in productivity without a large increase in cost.

The approach of assigning a talk path only when someone is speaking has been employed on the circuits between Hawaii and the mainland with a technique called time assignment speech interpolation (TASI). This technique was very successful in achieving an increase in occupancy on extremely expensive circuits that was not noticeable to the customer.

Packet switching can pipeline the various connections into whatever facilities are available and let the receiving end determine the sequence. The evolution to this form of transmission to information forms other than data will be slow as the various problems are worked out. It is quite likely that switching of voice and data calls within the same pipeline will soon be offered by LANs.

Packet networks originally boasted about their ability to route packets in any sequence and have the destination reassemble the original sequence. Present packet networks use a *virtual connection* where packets from the same session are routed over the same route unless unusual congestion or failure occurs. However, the packet still uses the connection only when information associated with that session is present. This scheme reduces the overhead necessary to reassemble the packets at the destination node. It also takes packet networks one step closer to circuit networks. The principle difference between a voice or circuit network and a data or packet network is that a packet network, during congestion periods, allows variable delay in the transport of information, whereas a circuit network must have a fixed delay for a voice connection.

Packet networks are now firmly established for numerous applications, although most of these applications are localized for a particular function. For LANs to expand, standards for interfacing to and between data networks are necessary. These standards also would give impetus to the various multiplexers and concentrators that can be connected to data networks.

6.2 X . . . Series

Numerous standards in data networks have been invented, and a need exists to establish interfacing techniques among these systems. A plethora of manufacturers for multiplexers, terminals, and modems need standards to direct their developments to the broader market. A continuous and efficient international standardization activity is indispensable in the dynamic world of data communication. Fortunately, an international committee (CCITT) is working toward this goal. It has developed some standards for interfaces and protocols between the various elements of data network. These recommendations normally are documented in the X . . . series. Some of these standards are shown in Figure 6.2.

The X.25 standard is the most common protocol, because it represents the interface to the data network from either a computer or a terminal, and many manufacturers talk about their products' X.25 compatibility. The packet assembly/disassembly (PAD) represents the device that prepares the packets for transmission to the destination according to the standards of the network. It also interfaces with the terminal for breaking

Fig. 6.2. X . . .
Series Interface
Standards

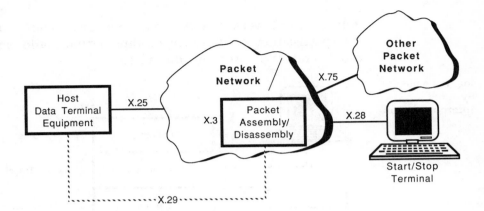

the packets into usable information. When a PAD is required, the X.28 standard applies at the PAD end, and an X.29 standard applies for interface between the distant end and the PAD. The X.3 standard will specify the PAD in the data network. The X.75 standard allows packet networks to communicate with one another on either a domestic or an international basis. Figure 6.2 shows the interworking of these standards.

These standards have brought some order to what could be an uncontrolled situation, and the members of the CCITT are to be commended for what must have been a painstaking task.

Two other X . . . standards are of interest. The X.121 standard is a recommended numbering plan for a public data network, and the X.200 standard is a protocol reference model of the open systems interconnection (OSI). The OSI breaks the connection into seven levels or layers from the individual terminal accessing the network to the physical media used in the network. The advantage of this is that incompatibility may exist across the network; however, that incompatibility may only be at one of the layers and readily corrected.

The clear delineation of these layers has permitted application or interface processors to be built, allowing communication between networks or between networks and host computers. This area of communication is where the data networks truly shine. These standards will not instantly allow a host computer to communicate with the world, but they eventually will prevent that host from becoming enslaved by its own protocol.

Figure 6.3 is a look at the OSI. The first three layers deal with routing the call through the network and are defined by the X.25 standard. The other four layers are concerned with the interface for the session itself. The standards for these layers presently vary from network to network. However, each network manufacturer has defined the specification for these layers, thus permitting interface from interested vendors or protocol converters.

Fig. 6.3. Open Systems Interconnection

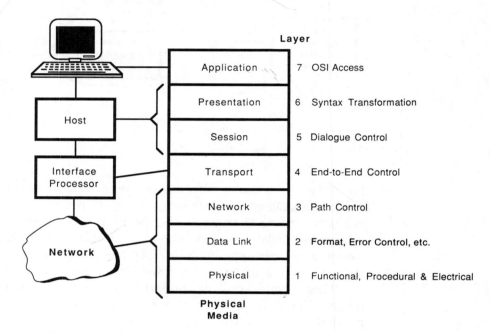

6.3 Data Networks Applications

A few months ago I was attempting to move my family to Florida for a vacation. I called Delta Airlines for reservations. Not only did the employee help me find an airplane to accommodate my group, but she was very helpful with the hotel reservations. She informed me that "The following hotels have available rooms in that area, and the rates are...." This information permitted me to make all my reservations in one call. I pictured her scrolling through the screens on her terminal to provide me with information received from a computer that was probably 1000 miles away.

Most of us have used a similar data network when reserving a flight (or flights) from anywhere in the United States. The nice thing about this airline system is the interface between the employee and the customer.

Although there are a myriad of data networks in use, few are as service oriented as the reservation networks of the airlines or the *Wall Street Journal* network. Many people are still amazed that *The Wall Street*

Journal can show up at home or the office every day when it takes four days to move a letter across the city.

Data networks began as a method for computers to communicate with one another, but they have branched out to become an integral part of service from a business to its customers.

Those data networks known as real-time systems have created the greatest growth in data communication. The airline reservation system is a prime example of a real-time system; that is, the person at the terminal is not only able to obtain the information on flights and hotels but is also capable of updating the data base when a reservation is made. The banking industry is also a big user of real-time systems so that they can update accounts as rapidly as possible. In the future when I make reservations, my bank account and the reservation data base will be updated at the same time.

Other applications of data communication systems include methods to move information on branch or satellite operations to a central data base (on-line/off-line data transmission), methods to make the information in a computer available to a user without updating (inquiry/response systems), and batch processing operations.

Data networks can provide information for video displays, printers, large batch operation, or computer-to-computer communication, and can vary in speed from 300 b/s (actually some terminals operate at lower speeds) to 2 Mb/s. For speeds of 300, 1200, and 2400 b/s, a dial-up modem connected to the public long-distance network will provide good data capability for the infrequent user. The prevalence of the public switched-phone network is a major advantage of this approach, because it only requires a modem at its extremities to get started.

As the data requirements or range of applications increases, we will want to investigate special data communication or separate data networks. We will want to look at multiplexers and concentrators for the terminals. A multiplexer or concentrator will take a group of slow-speed terminals and interleave them to a higher-performance link to the host computer. Today, there is little difference between a multiplexer and a concentrator; both are controlled by programmable processors. A multiplexer used to be a "dumb" version of a concentrator. These terminals can be located in the same building as the host or remote from it. In either case we are starting to put together a data network, because we have the main ingredients: terminals, modems, multiplexer/concentrators, and a main computer that these devices would like to access. We can continue to use the public voice network, but our occupancy has increased sufficiently to justify some leased lines, some wideband facilities if high-speed is necessary, and, possibly, a look at data networks.

6.4 Types of Data Networks

There are many data networks, and they are all different. Numerous combinations can be derived from these arrangements, but, for practical reasons, data networks will be categorized as general (Telenet), specific (SNA), LANs, and special (multiplexing). When a business decides it has outgrown modems and wants to move up to data networks, it needs a way to solve any compatibility problems with distant devices either inside or outside the business.

General Data Networks

An example of a general network is the GTE Telenet, which interfaces with different devices, operates within the international X.25 standard, and provides synchronous or asynchronous terminal support. Telenet is a packet-switching network that uses high-capacity transmission lines and electronic switches at its nodes to interconnect various facilities. If we want to interface a teletypewriter with an ASCII protocol, we hook a teletype unit to the network. If we want to run a program from a distant IBM host computer on our terminal, we hook our terminal to the network and run the program. In other words, Telenet can provide the data communicating equipment (DCE) for most equipment and, thereby, interface the equipment to the customer's computer or any computer on the network to which the customer is permitted access. The structure of the Telenet system is shown in Figure 6.4.

Systems Network Architecture

The IBM systems network architecture (SNA) is a specification established to allow communication between IBM products and between vendors and IBM products. It is an excellent example of a data network built for a particular problem.

The principle purpose of the SNA is to permit any terminal user with an IBM or IBM-compatible device to communicate with any service or application within the network. IBM and other vendors offer network interfaces and converters to permit an SNA to communicate with other SNA and non-SNA data configurations.

A configuration interconnecting terminals, processors, links, and so on, that meets this specification is called an SNA network. It is a user-

**Fig. 6.4. Telenet
Data Network**

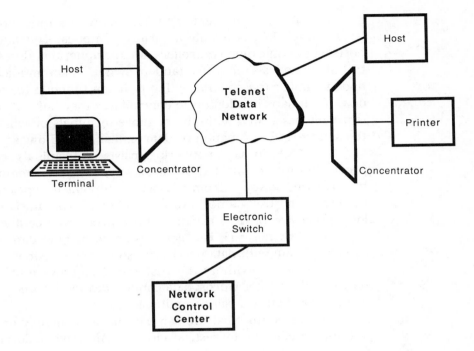

application network that has been continuously enhanced as new needs emerged. It is tailored to the network owner requirements to provide the capability needed by a business or service. An SNA network gives users a broad range of choices in telecommunication services and facilities.

The major thrust of SNA is to provide compatibility through the entire network from terminal to terminal or from terminal to application unit. Most networks will transfer the call or packet to a particular destination, and then it is a question of whether the end users can communicate. This communication is accomplished by the use of layers within SNA. The application layer is where the work is performed by the end-user. The function management layer is where the data flow is controlled and messages are transformed, if necessary, for presentation to the end-user. Also, requests for file access or transaction are handled here. A transmission layer is responsible for routing requests and responses throughout the network.

This approach is quite similar to a telephony switching concept where the functions are divided into the phone user, the switching equipment, and the transmission between switching units. Versions of SNA are available that support different operating systems and software control programs, depending on the customer needs and the existing equipment.

The SNA is quite similar to telephony. The functional management layer within SNA is mainly a variety of nodes. The network can have hosts, communication controllers, and peripheral nodes. Host nodes control other nodes similar to a tandem switch. A communication controller node is similar to a toll office; that is, it can function to control terminals or workstations in addition to controlling other links. A peripheral node is usually associated with one or more terminals or workstations and is the source and destination of data similar to end offices.

The SNA physical network consists of terminals, controllers, and processors interconnected by various types of telecommunication links. In this case, however, from a user's standpoint it appears to be a single network. The user is not aware whether a program is being run on a local processor or on a remote central processing center.

The SNA network is a good example of how data communication can coordinate with data processing so that the customer's various tasks are accomplished without the customer being aware of where they are performed. In other words, data communication is rapidly becoming part of this paradigm we call computer systems.

To achieve this goal, we can use circuit switching or packet switching and satellite or terrestrial circuits. Although tailored to particular needs, a general structure of an SNA network would be as shown in Figure 6.5.

**Fig. 6.5.
Overview of
SNA**

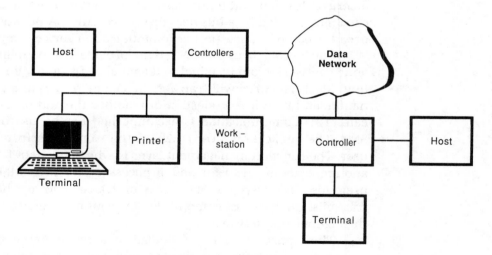

SNA networks have grown to a point where it may be economical to interconnect different SNA networks via gateways. A gateway is a facility used to interconnect networks by accepting messages from one network, translating them to a form understood by the other network, and routing the message to its destination. Most gateways only interconnect networks that are operating on the same protocol. For example, an SNA gateway would interconnect two SNA networks but would not interconnect an SNA network with an Advanced Research Projects Agency (ARPA) network or a public data network that uses an X.25 data scheme (with an X.75 interface and an X.121 addressing). Gateways for interconnecting networks with different protocols are the goal of the CCITT interface standard X.75. This approach uses the global addressing scheme X.121. As data processing expands and the trend to decentralization continues, more gateways will be necessary between these networks, and the importance of standards will be apparent.

Let's look at some variations on these networks and see how data communication handles various requirements. The primary example would be a large corporation using a data communication network to centralize its data processing equipment. If most of its data processing is done with IBM equipment, the company would use a network similar to the SNA. Where there is a variety of equipment, some interface devices to an SNA network may be the answer, or access to a generic network, similar to Telenet, may solve the problem. Digital Equipment Corporation recently announced that it is marketing an interface from its DECNET to an IBM computer. Also, application processors are being designed to allow more interfaces as data networks spread and the need for communication between them increases.

Digital Data System

The Bell System also provides a data network known as digital data system or Dataphone digital service (DDS), where the data transmission is maintained in a digital format as it progresses through the current circuit network. This is accomplished with a four-wire, full-duplex line running from the central office to the user. At the central office the signal is multiplexed onto a T1 facility for connecting through the system. It is often necessary to share the T1 line with voice traffic and data under voice (DUV), a very clever technique that allows the voice and data to coexist on the same circuit. The full cost of the facility is shared between the voice traffic and the data traffic, thereby keeping the expense to a minimum.

AT&T has divided the country into digital serving areas (DSA) authorized by the FCC. Most areas are metropolitan. The users can have a rate up to 56 kb/s, but the bulk are at 4.8 and 9.6 kb/s.

6.5 Local Area Networks

I once became involved in a heated discussion among several technical people arguing about how to make two computers located at different ends of the country communicate with one another. One group argued that the telecommunication network was at fault because it couldn't provide interface between the two computers. I asked if the two computers could communicate with each other if they were in the same room. After much discussion, it was agreed they couldn't.

The telephone network can no more help two incompatible computers to communicate than it can help (right now) two people who speak different languages to communicate. However, systems known as LANs have emerged to assist computer and computer-related equipment to communicate with each other. A LAN allows the computers or electronic equipment located in the same room, building, or campus to communicate with each other.

The best way to picture a LAN is to assume we have some workstations (word processors, CRTs, or PCs), and we want access to a printer, disk storage, mainframe, or another workstation. However, we only want one printer and disk storage for all the workstations. A LAN allows us to communicate with the printer, the storage, or another workstation without going to the mainframe.

The LAN is far more than advanced information, transmission, and processing facilities. Sharing of resources has been part of the telecommunication philosophy, which has built a mathematical science to accomplish this sharing. LANs are based on availability of certain common resources and on the distributed processing capability of modern electronics. The LAN also has contributed to the productivity of the office by allowing groups of people to function efficiently in a well-integrated manner. The most common use of LANs is PC networks. As these PCs find their way to more and more desks, demand to expand the office infrastructure increases. LANs are responding to these demands. But the really neat thing about a LAN is its ability to satisfy the nonprogrammer professional—the end-user—who wants to press a mouse when the cursor points to a printer or to press directly a printer symbol on the screen and pass the document to a printer without learning a programming language. Figure 6.6 shows the basic LAN configuration.

One generally accepted definition of LAN describes an information transport system for data transfer between terminals and resources (such as printers) and indicates that the only difference between a LAN and other data networks is that LANs serve a limited geographic area

Fig. 6.6. LAN Communication Structure

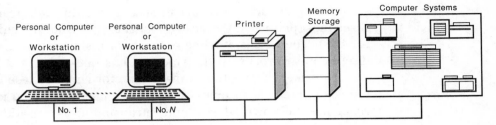

while other networks have no such limit. I disagree with this definition because it implies that a LAN has a limitation on how far it can transmit, and though that may be true today I doubt it will be true in the future. A LAN is another data network, perhaps misnamed, using distributed processing and presently confined to a limited geographic area due to the techniques used. A better definition would describe a LAN as a data network that does not use a common carrier facility for transport.

Another reason for the emergence of LANs is the need to pass data around a local area rapidly and easily with a simple system. Because as much as 80% of the information an individual uses typically can be derived locally, development of a LAN to solve the local data transmission problem appears to be a natural extension of current technology.

Does that mean that LANs will disappear when the PBX or mainframe has good data-switching capability? Probably not. The LAN will be able to operate in certain local areas at a much lower cost than the PBX or mainframe, and LANs are becoming deeply entrenched. In addition, PBXs may take on a dual role insofar as they may act as a LAN or as a device for internetworking LANs. These functions must be considered in any future PABX (private automatic branch exchange) design.

How It Works

The basic function of the LAN is the exchange of information between users or between users and servers in the network. The information is transmitted so that everyone on the network can see it, but is coded for the particular destination. The analogy I always think of is the one from early telephony. Early party service had what was called 10-party service with coded ringing; that is, all 10 lines would hear the ringing, but only one person would recognize it as destined for his or her phone and would pick up the call. LANs operate the same way, although they are also capable, in some cases, of transmitting messages without getting an answer recognition.

There are different ways to accomplish this data transmission. The three most popular LAN arrangements are the star, the ring, and the tree (or bus). The PBX is a prime example of a star network; therefore, this discussion will focus on the ring and bus networks.

Figure 6.7 shows the configuration for ring and tree LANs. The ring topology arranges all points via links (sometimes called nodes) that can recognize their own address and retransmit messages addressed to other points. The messages and controls are moving around the network from point to point, which will start to introduce delays as the network grows. The controls of the network tell each point on the network when it can transmit (deterministic), so there is normally only one message on the network at a time. A plethora of variations exist for the control and ring structure to gain more speed or reliability, but the principle of this access method remains the same.

The token-passing version of ring topology recently obtained a major endorsement from IBM when "Big Blue" accepted this configuration as its standard LAN. This obviously will cause many companies to introduce token-passing ring LANs.

The tree or bus structure access method allows each point to transmit whenever the bus is free (statistical). All points will see the address of the message, but only the destination point will accept the message. Again, as the network grows, the chances for collision (multiple points attempting to transmit at the same time) increase. Most of the current work on this type of structure focuses on this contention problem. This structure is a forerunner for acceptance. It is Ethernet structure and used by Xerox, DEC, and Intel. IEEE-802 standard is based on this structure.

The transmission media used for these networks is important, because it could establish standards for future telecommunications. Again, different media are used, but only coaxial cable and fiber optics are discussed here. Actually, the coax cable has a baseband entry and a wideband (sometimes called broadband) entry. The baseband is used for Ethernet and other networks and operates like many data applications, generally an on/off keyed digital data structure. The broadband media modulates the voice, data, or video information it carries. It uses the same FDM scheme as cable TV.

Fiber optics may offer the greatest future as a LAN transmission media because of its enormous bandwidth capability, small size and weight, immunity from electrical interference, and eventual low cost. Presently, the advantage coax cable has over fiber optics is its ubiquity and its ability to move any terminal to any point on the network.

When considering installation of a LAN in a building, we need a wiring strategy that is sufficiently flexible for reconfiguration and expan-

**Fig. 6.7. Ring
and Tree LANs**

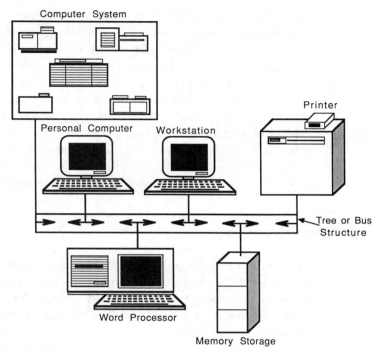

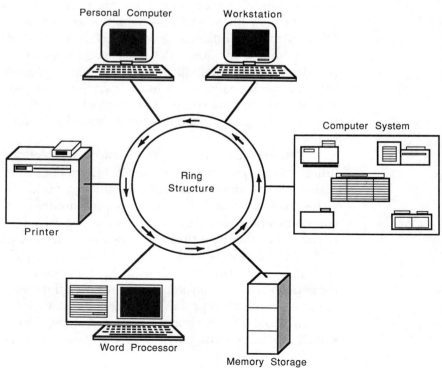

sion. The wiring scheme also must be reliable, available, and serviceable. A vendor should be able to demonstrate some wiring plan associated with a system. The evaluation of this aspect of the system could be critical, especially in handling expansion. There is nothing worse than rewiring an office ceiling.

With more than 200 companies involved in supplying LANs and others entering almost daily, we might ask whether they will remain after the telecommunication network has implemented the same features. I feel there will always be a need for local networks because of convenience or security considerations. The problem will be the interface of LANs when internetworking is attempted.

Internetworking

The present growth phase of LANs relates to the internetworking of these networks. This phase was brought about by the independent growth of LANs in the office, the factory, and the development area. The ability to pass information from one of these areas to another not only increases productivity but also improves the time factor and reduces errors.

The LAN for any one of these applications probably was purchased from a single vendor who manufactured both the terminal and the resources, so there was no compatibility problem. Also, a LAN that is efficient for the office is different from a LAN that is efficient for the factory, so the various LANs most likely have interface and/or protocol problems. It is not helpful for the communication manager and the information manager to sit in their chairs pointing fingers at each other and saying, "I told you to use the common carrier (mainframe)." Internetworking is a problem to be solved via the PBX, mainframe, or both.

The problem can be divided into a connection problem and a communication problem. The connection problem may be solvable via the digital PBX as voice/data find a common ground via these switches. For years we have called these switches "communication systems," although communication can only occur when the two parties speak the same language. In this case these switches are interconnection devices, which is the first step in solving our communication problem.

The other problems involve transmission media and protocol. If these problems are solvable (in many cases they may not be), the solution lies in an application processor, which translates one language to another.

In dealing with vendors who are selling these application processors, customers must make sure they are not only getting a device to allow communication between LANs but one that will give them the speed they need. Many applications require only minimal speed, but if the office area wants to use the laser printer in the development area for large documents four times a day, speed is a major consideration. Although the vendor may quote speeds of 1.2, or 10 Mb/s, the resulting speed (throughput) after the interface may be substantially different.

Although LANs are enjoying tremendous popularity, they are going through "an awkward phase." There is a need for standards because almost all standards today are de facto, and if internetworking is to work, the number of combinations must be reduced to a manageable size. LANs now are poised somewhere between being an integral part of the telecommunication scene and being a convenient method for accessing special devices.

Market View

The basic market for LANs has been the large organizations—businesses, universities, or laboratories—that require interface to the various information processing systems. As such, LANs are required to provide high-speed communication and adapt to the standards of the information processing systems. LANs meet these requirements by using communication techniques and transmission media different from those used by telecommunication vendors. In other words, a niche, created by the telecommunication and computer vendors' inattention to the communication link between the user and information processing, was filled by LAN vendors.

The problem with the niche for LAN is that information processing and telecommunication vendors are about to wage a war in this battlefield. If it isn't enough to have IBM and AT&T battling over a market niche, the data networks also have their eyes set on interfacing with terminals and, thereby, bypassing LANs.

The LANs' vendors have been attempting to keep the telecommunication vendors away by saying the PBX and the LAN serve different markets. The PBX, they claim, serves a low-speed, point-to-point communication market that is different from the high-speed, protocol-dependent, information processing market of a LAN. They insist that even the cabling for the two systems is different, with PBXs requiring only voice-grade twisted pair, whereas a LAN requires at least a coaxial cable.

These various arguments recently disappeared as IBM announced their LAN and stated, quite clearly, it would operate with voice-grade twisted pair. This announcement, in essence, stated that the niche for

LANs will disappear, and a communication scheme with a data network (SNA for IBM) or a voice network will be used in the future.

Data networks and voice networks, with the demise of the hierarchical switching scheme, rapidly are becoming two-tier structures, and it appears that IBM's structure also will follow this path. IBM announced a 4-Mb/s token-passing LAN with current interface to its System/370 mainframe and its S/1 minicomputers. However, IBM is expected to add 16 Mb/s for high-speed devices and SNA. The announcement also spoke of the ability to operate with twisted pair. IBM architecture looks similar to other two-tier structures except that mainframe computer and other high-speed units are connected to the upper tier and PCs, workstations, and other slow-speed devices are connected to the lower tier. It would not be surprising to see a PBX interconnect at this tier, thereby eliminating the current myth that LANs and PBXs are different solutions to different problems.

A PBX vendor had better be looking at gateways to the higher tier in order to stay in the marketplace. The strategy for the PBX is conspicuously complicated by IBM's announcement, with little information forthcoming from AT&T. Of course, AT&T is busy with their own announcements concerning new electric typewriters and new computers. These giants of telecommunication apparently have adopted a revolving-door strategy reminiscent of the one used by France and Germany during the first month of World War I—attack the other guy's flank. Unfortunately, in the next few years the battle may be in the trenches.

Application of LANs

In spite of this uncertainty the requirements still exist for a LAN within an organization, and the application can be satisfied with cost-effective equipment. An initial method for determining whether a LAN is required is to contact the groups within the organization who would have interest in interconnecting terminals or sharing resources. A survey can be used to provide information on type, speed, and use of terminals. In addition, it would provide information on whether interface to the mainframe is a requirement and on the type of resources needed to interface.

The information gained from the survey should help determine which type of LAN will best serve the requirements. However, the following items must be addressed before the LAN is selected:

- System configuration
- System operation and application

- Capabilities
- Physical requirements
- Cost of equipment
- Cost of support
- Time frame for delivery

It is tempting to hire a consultant for this evaluation, and help can indeed be obtained from someone in daily contact with a rapidly changing technology. However, that doesn't exempt you or me from learning as much as possible about the selected product and training the individuals who will use the system.

PBX and LAN Comparison

One method for expanding the LAN is by interconnecting the LANs with other networks, especially packet networks and long-distance networks. This step, although logical, will create a clash between the PBX market and the LAN market.

A LAN, by definition, can share resources or data from several terminals or leave messages for other terminals, depending on the application. Packet and long-distance networks are used for communicating information (voice or data) between users. However, numerous applications require that these networks be tied together. The possibilities are

- LAN to LAN
- LAN to packet network
- LAN to long-distance network
- Packet network to long-distance network
- Packet network to packet network

The fact that only the long-distance network has established standards for interfacing will create some complex topology when these other networks are interfaced. This situation is further compounded because LANs have simple message and acknowledgment procedures, and any network interface must have a complicated protocol to ensure valid transmission. Networking is the key to the future: the recent announcements by IBM and AT&T validate that statement. IBM has recently placed their seal of approval on LANs but added to that endorsement a networking concept for interconnecting their LANs with the mainframe.

Packet networks can be long-distance networks, although the term "long-distance" network normally refers to the voice network that has been around since the 1930s. A packet network is implemented over leased facilities using virtual circuit connections. The voice long-distance network establishes a connection between two subscribers via a distinct path, and all circuits associated with that part are permanently assigned as long as the connection is maintained. With packet switching the end-to-end connection (a logical circuit) is known, but the links between the two ends are dynamically assigned when a packet is being transmitted. Since the packets can have variable delays in reaching the destination, the packet network is not considered suitable for voice communications.

The interconnection of PBXs and LANs or handling of both functions by a single unit is the challenge presently facing network designers. The physical connection is different for voice and terminal, the method of communicating is definitely different, and the architecture for LANs has unique approaches not seen in digital PBXs.

Table 6.1 illustrates some of the visible differences between the two approaches.

Table 6.1. PBX/LAN Telecommunication Structure

Feature	LAN	PBX
Architecture	Ring or bus	Typically star with hierarchy
Bandwidth	Up to 10 Mb/s	Up to 64 kb/s with 1.54 Mb/s
Routing	Simple and based on architecture	Extremely complex due to distance and number of customers
Interface standards	Normally requires an application processor	Uses well-defined industry standards

As the information presented in Table 6.1 indicates, LANs were developed to solve unique problems without being concerned with interfaces, routing, or protocols. A LAN attached to a data or long-distance network can access useful resources such as electronic bulletin boards, but the access will have an impact on the capability of the system. If someone spent all day searching a data base, the other users of the LAN will find the response time deteriorating, and the usefulness of the unit will be questioned. The LANs were originally conceived to solve problems for the users that the mainframe wanted millions of dollars and an infinite time to implement. If a LAN is interfaced to a long-distance network, the user wants the best of both worlds—the convenience and speed of the LAN plus access to additional information or resources. In addition, the cost should be reasonable.

The network access approach will probably involve a "layering" of the various systems. In other words, the LAN, as long as it is communicating with its users, will enjoy the speed and responsiveness it presently enjoys but can choose to route through another tier or layer for special access. The latter service will be obtained via a gateway to other networks with little effect on the operating speed of the LAN.

Some technical comparisons between LANs and PBXs are shown in Table 6.2.

**Table 6.2.
LAN/PBX
Comparison**

Requirement	PBX	LAN
Interfaces for data processing and telecommunication	Digital architecture with SPC for data exchange	Bus structure with microprocessor controls
Interface for voice/data workstation	Twisted pair with dual bus arrangement	Twisted pair with single bus for voice/data
Long-distance high-speed data	Interface to packet network	Need interface to packet switch
Common carrier interface	Normally available with switch	Interface development required
Video interface	Probably a separate switch	Broadband is capable but not available with twisted pair

Future of LANs

LANs are typically associated with a limited amount of switching, because their primary responsibility is resource sharing. For various types of LANs, whether star, ring, or bus structure, the routing is basically fixed. This permits additional growth for this technology as more applications are developed for resource sharing. One approach to this is the voice/data switch.

The advantage of the voice/data switch is that many offices are finding a need to switch both types of calls during the workday, and they would like one instrument on the desk to handle both connections. The voice/data switch must be capable of switching at a speed of 56 kb/s or 64 kb/s to eventually interface to the ISDN (integrated service digital network), but will initially present problems interfacing to 19.2-kb/s devices, typical in modern data networks, and asynchronous terminals. Arrangements may be possible for these devices, but the main advantage is the increased number of calls and terminations the switch can handle.

There are other advantages, not the least of which is integrated call accounting, which is important in virtually any business. The combined switch can also optimize routing for the calls via MERS (most economical routing system) or equivalent programs. The routing of the call over the network will expand the world of the LAN and permit the connection to be carried over the 1.544-Mb/s carrier.

Local area networks use packet data for their format, which is consistent with a national data network; however, the content within each packet and the envelope for the packet are not readily convertible to the format of the national data networks. However, the advantages of shared facilities will force a solution to these problems.

The future of a LAN was tremendously aided with the IBM announcement regarding their standard and their encouragement to other vendors in interface as well. This strategy will probably do for the LAN market what the introduction of IBM's PC did for that market. Remember, however, we are starting to see a number of fallouts in the PC market and the same can be expected for the LAN market. At the customer's level the information processing and the voice processing will be handled by one switch; whether this switch is a LAN or a PBX or a combination switch is about to be decided.

6.6 Multiplex or Concentrator Networking

Originally the interface to the packet network was via DCE which would interface the various DTEs with the network by meeting the requirements of the network's vendor. With the protocol firmly established for the various interfaces and networks, a variety of concentrators with multiple inputs can be purchased. An example is shown in Figure 6.8.

The aforementioned data unit normally has its own processor and buffer. This unit processes the necessary diagnostic and reload capability consistent with the customer's requirements. Typically, the interface to the network is 56 kb/s, and the number of links required can be determined by the number and speed of the devices connected to the concentrator.

We need to transmit billions of bits of information almost instantaneously for this new age, yet the number of channels available for this transmission is limited, so some type of allocation system must be used to assign these channels. Current technology has engineered fiber-optic links capable of transmitting billions of bits of information (a billion bits is known as a gigabit) over great distances. Independent channels making

Fig. 6.8. Various Interfaces for Data Equipment

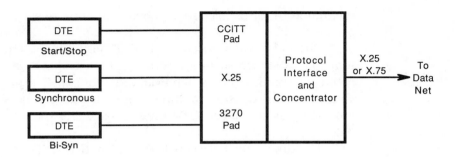

different demands on either high-speed channels (e.g., fiber optics) or on limited cable pairs in a congested area leads to an allocation scheme known as *multiplexing*.

Multiplexers

Multiplexing is the technique for allocating several low-occupancy calls onto a high-occupancy channel. It can be accomplished with FDM or TDM but the end result is basically the same.

The multiplexer was designed as part of the telephone industry's efforts to have a number of conversations or messages share one transmission facility. The computer industry adopted multiplexers to allow more data devices to share pooled modem resources. For both industries the multiplexer has become an integral part of any strategy to provide networking and service at the lowest cost. The multiplexer is an effective way of bridging the gap between a slow-speed terminal and a high-speed line.

The actual occupancy between a terminal and a computer or from a link to a computer is extremely low in terms of traffic, and the cost of running cable between every device and the main frame is prohibitive. The multiplexer or concentrator is an excellent answer to this problem and is designed to operate between the terminals and the serving device. The basic idea behind a multiplexer is to gain additional bandwidth capability from each channel within a transmission media without affecting the service to the user. The scheme also avoids the need for an inordinate amount of cabling.

Until recently, frequency-division multiplexers were more practical for this application, but time-division multiplexers have now taken the lead. In TDM the channels share the transmission by "taking turns," each being connected to the line very briefly, then replaced by the next. This technique requires precise synchronization between transmitter and receiver.

The most widely used method of multiplexing within the computer industry is statistical multiplexing, which requires even more sophisticated electronics but is achievable at competitive prices with modern circuitry. Time-division multiplexing dedicates a time slot to each terminal, whereas statistical TDM only assigns time slots to the active terminals.

The difference between TDM and statistical TDM is shown in Figure 6.9.

Fig. 6.9. Time-Division and Statistical Multiplexers

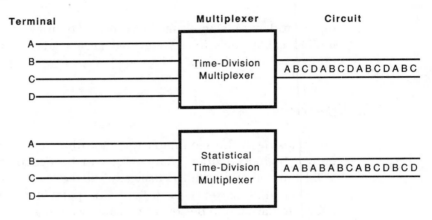

To increase channel utilization, we can use statistical multiplexing or asynchronous TDM for communication. The basic idea is to switch from one user to another whenever one is idle and the other is ready to transmit data. With such an arrangement each user is granted access to the communication channel whenever a message is to be transmitted.

Statistical multiplexers normally are arranged with memory and controls to detect the "busy" state of a terminal, assign a time slot, pass information to the receiving end, and provide a buffer memory.

For most nonvoice applications the statistical multiplexer is ideal because it provides a vast improvement in circuit efficiency while maintaining the service level for the user. Most voice applications will continue to use the time-division multiplexers until an answer is found to the variable-delay problem.

A form of statistical multiplexing for voice communication is being incorporated into digital switches (e.g., Bell's SLC-96). This multiplexer normally is equipped with 24 channels to the switching center but can serve more than 24 customers up to about 100. Channels are assigned to the customers as demand arises, and there is a good chance that no more than 24 channels will be required. However, if more than 24 conversations are needed, then additional requests are denied, whereas in the statistical multiplexer they are delayed.

Another variation on this is TASI equipment, used on transoceanic cables to assign a customer a channel only during active conversation. TASI is designed to detect a customer's speech and assign a channel during the speech process. During the idle time no channel is assigned. TASI was probably the first statistical multiplexer and, due to the falling cost of the electronics, this technique could play a role in the forthcoming integrated voice/data market.

Configurations

Multiplexers can be connected to terminals in several ways and are becoming a continuation of switching as features are added. The most common configuration is the point-to-point system shown in Figure 6.10.

**Fig. 6.10.
Concentration
of Low-Speed
Devices**

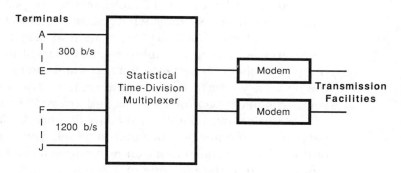

The requirement is to connect a number of colocated terminals to one or more computers located at a distant point. Point-to-point systems save in modems and links charges and are extremely beneficial when many terminals are colocated. Office areas with numerous terminals are ideal for this type of application because there are other multiplexers available if one of the units fails. This is a common objection to point-to-point systems.

For users needing greater reliability or for larger applications, the nodal configuration in Figure 6.11 can be used.

These systems provide added reliability or improved service criteria by terminating the users on both multiplexers and sharing the load during normal operation. If one multiplexer fails, the other will carry the

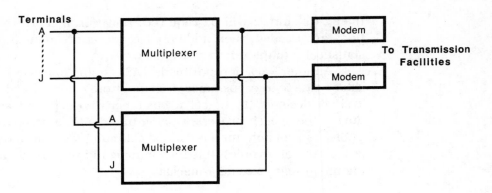

**Fig. 6.11.
Reliability
Arrangement of
Multiplexers**

load. One warning for this arrangement is that the links should be run on separate routes to provide additional reliability.

Network processors associated with current multiplexers allow a user to build a multinode configuration with any level of reliability or load balancing desired. These networks can provide many advantages, especially when there are many terminals.

Other configurations of multiplexers are possible, including multipoint or multidrop systems where several multiplexers are arranged on one link and are polled for information at certain intervals to prevemt imterference. A good multiplexer is transparent to the user, reduces the number of channels between the users and the mainframe, and provides the necessary interface for communication. The interfaces are based on standards developed for modems and are readily available.

Statistical multiplexers are being arranged with controls to allow a particular computer or other unit to be selected. This is obviously a view of the future for the multiplexer as it enjoys more functions and becomes more critical in the planning of a network.

6.7 Terminal Interface on Data Networks

Data Terminals

Most data processing operations have dedicated or leased links interfacing data communication facilities currently in place. Money is saved if these links could be consolidated and switched through a communication system without any service impairment. Service refers to a transparent connection and the ability to handle the necessary bandwidth. Some examples of data processing connections are point-to-point on-net, point-to-point off-net, and multidrop. Common terms for data processing network connections are DTE and DCE. Recall that DTE refers to the unit attempting to communicate (a mainframe, a terminal), and DCE communicates with the network.

For the point-to-point on-net call the arrangement before switching is shown in Figure 6.12.

Fig. 6.12. Point-to-Point On-Net Call

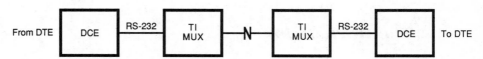

Data processing equipment can be owned by the customer and need not be leased from the telephone company. A saving is realized, even though the connection is not switched, because the T1 unit can be part of the network. Additional savings are possible if the connection is switched, which will be discussed shortly.

For off-net connections the arrangement can be as shown in Figure 6.13.

Fig. 6.13. Off-Net Connection

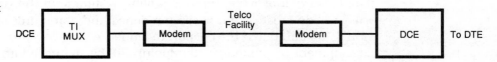

For a multidrop arrangement a digital bridge is used at points where the devices are located. Figure 6.14 shows one possible configuration of what has become extremely popular.

Fig. 6.14. Multi-drop Arrangement

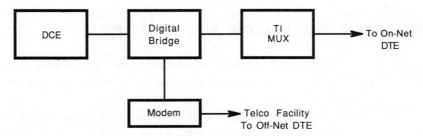

Eventually, switching systems will have the ability to switch these devices, so of these facilities will be increased.

A typical operation with a switching system is the introduction of two distinct structures with a common interface to the customer. The customer will initiate a call, and the system will serve it and determine whether a data or voice operation is necessary. Depending on the decision, a particular interface to provide the correct functions will be associated with the connection. For data calls interfacing with DCE a packet arrangement is used, thereby providing the customer with a minipacket network in addition to the voice network. The packet network incorporates a packet manager for interfacing the various arrangements.

The Workstation

Although most desks are equipped with telephones and will continue to function quite well with that instrument, there is a growing demand for instruments that can access the computer, data, as well as serve as a phone. The data can be facsimile or electronic mail. The impetus for this workstation (for lack of a better name) is to eliminate or greatly reduce paper flow within the office and to increase communication for the user.

Before tackling the problem of introducing the workstation into an established environment, we should discuss the advantage of the workstation. For the user the workstation will become the office. The various data bases will provide the user a filing system, and the voice messaging will provide most of the correspondence and communication with individuals up and down the corporate ladder. Access to status charts from a development or monetary standpoint will be the user information channel. One advantage is that the user can ask stupid questions at the workstation and not be embarrassed by the answer. This is sometimes called a friendly interface.

6.8 Data Network Planning

The arena of data communication appears to be quite technical, especially when newcomers are faced with constantly changing jargon and equipment.

Users must be able to distinguish between something state-of-the-art and something that meets individual needs. Someone who understands data processing will not necessarily understand data communication. Someone who is an authority on voice communication may not be well versed in data communication. Voice communication is based on person-to-person service with well-defined characteristics. Data communication seems to encompass all the jargon of computer science and telephony. However, each discipline is separate and distinct. By setting guidelines and goals that will target this technology to the end-user (the person making airline or hotel reservations, the engineer, the accountant), we can achieve a systematic approach.

Voice and data networks are coming together partly as a result of the end-user's demand for one instrument on the desk. Voice traffic will account for the largest volume on the voice/data network. However, data and data communication features will continue to have the greatest growth as users recognize the broad range of applications available with data.

Incorporating the Packet Network

A packet network can distinguish between various groups of users and can provide different bandwidths or data speed, depending on the requirements. This is a major difference between the circuit network and the packet network and is one of the major features of the data network that ISDN will bring to the digital switching systems. Future switches will distinguish between groups of customers and provide special facilities, routing, and circuits, depending on their needs.

Packet switching typically has its own bus structure, which is accessed differently than the circuit-switched network. Thus the switching systems of ISDN will need to separate the call to one of the networks after it has determined the type of service the customer is requesting.

For a business the switch must provide access to

- ISDN
- LAN
- Public packet network
- Public voice network
- Private voice network or special services
- Integrated phone and/or keysystem

Most of these accesses have unique requirements and must be developed separately, thereby necessitating complex projects for the vendors competing in the marketplace.

Data Communication Planning

Data communication planning has been difficult due to the rapid growth, feature demand, and regulatory environment. Data communication capabilities will continue to evolve as the computer and the telephone merge and additional applications appear. Some of the present data applications are

- Telemetry, which was presumed dead but is rapidly rising and requires bandwidth to support 100 b/s
- Data, which uses the bandwidth of the telephone network for speeds of 1200 b/s, 2400 b/s, and occasionally 4800 b/s
- High-speed data, which operates from 9600 to 64,000 b/s for inquiry/response and electronic mail
- Bulk data transfer, normally between computers for files or large programs and using bandwidth from 64 kb/s to 1.544 Mb/s; this range will also be used for video
- Interconnecting LANs, which operate at speeds of 1 to 10 Mb/s

The problem is to effectively build a network to provide these services while maintaining a cost objective for the total communication package. If separate bandwidths were purchased for each of these services plus bandwidths for the voice communications, an economic nightmare would have been created. Current networks are not far from this configuration.

One solution to the problem is the ISDN, but until it fully arrives, services are needed and cost-effective answers must be found. Some proposals for solutions include a private-line data line with sufficient bandwidth to handle computer-to-computer transfer, a packet-switching network operating from 1200 b/s to 9.6 kb/s in-house communication systems including time-sharing options (TSOs) to mainframe for inquiry/response and LANs for local applications. About the time the local application is installed and working, someone decides that another communication problem has arisen and this new system doesn't have the answer.

Characteristics of Data Networks

The particular data networks (Telenet, SNA) are discussed elsewhere, but some explanation of their characteristics is given here. One of the most important characteristics is the ability to provide monitoring capability for the network, especially in private networks. Monitoring in data networks normally is over a separate channel at a low bit rate and includes monitor, control, and testing of the various components within the network.

The problem, or advantage if you prefer that approach, with service networks is their need to be centralized. With most of the computer facilities being disbursed for the workers' convenience it appears inappropriate to centralize the service for communication facilities. One local approach is to add telecommunication service to the computer service center, which is usually manned 24 hours per day. This should avoid any increase in manpower during the off-hours because the increase in workload is slight. Certain maintenance functions should be centralized because the maintenance expert must have access to complete information for diagnostics.

Another basic characteristic is the access to the packet network. This can be direct (bypass) or via a PBX that would have a packet-access feature. The former would replace any modem or other access the phone company has provided. This is a service rapidly growing because the regulatory environment currently favors this approach. However, bypass does not provide an evolutionary step toward the ISDN, whereas the PBX with its packet access may provide this step. The advantage of the PBX is that data and voice are starting to merge, which is the whole purpose of this exercise; if not, let's go back to voice communication in a regulated industry.

PBX access is possible with current technologies and provides a way to maintain an investment while the evolution toward an integrated network occurs. That is, a PBX with this feature is usable in the ISDN era, whereas bypass is a temporary solution.

The customer is presented with a plethora of reasons for employing separate data and voice networks, but the real reason may be the vendors' difficulty in merging these disciplines. It is easier to talk of an integrated network than it is to design and build one.

However, the customer is the one who is going to end up with a potpourri of networks and will need to untangle the mess when the communication costs get out of hand. To meet the communication requirements, we need a multifaceted approach to the demands and capabilities. It is true that digital networks can handle data and voice communications, but there is too much variation in bandwidth, traffic requirements, service levels, and local interface. To provide separate networks for each requirement will eventually create chaos and not take advantage of the traffic peak or the economics of combined facilities. The main advantage of the integrated network is that it controls the communication environment for the customer rather than letting the environment control the customer.

7

Telecommunication Services

In previous chapters we have considered the various problems associated with the transport of a voice or data call. In this chapter we consider some additional core problems—standards and special services not considered years ago but becoming pervasive features in the present environment.

The ability of telecommunication to handle particular services is related more to the definition of the service and the establishment of a standard for the service than it is to the transporting of the service across a network. Telecommunication is so far-reaching, and the list of peripheral material for these services is endless. In many cases the jury is still out on whether there is economic justification for some of the offerings, especially teleconferencing and videotex.

This chapter will allow you to become conversant with these services. Many of the services lack standards, which inhibits their ability to grow rapidly. It is worthwhile to cover the various organizations and groups that are diligently working on this problem for the industry.

7.1 Standards Organizations

In 1983, phone companies were convulsed by this terror known as deregulation, and predictions ranged from chaos to total destruction of the phone system. The ease with which the phone industry has adopted equal access, for example, as the first step toward a new network is very impressive, especially when one considers that for almost a hundred years, the network has depended on the Bell System for its standards. This is not to say there have not been problems, but most customers have perceived very little change in their phone service. This is fortunate, because many more changes are forthcoming. The institution we know as telecommunication service is destined to become information service and the structure of the vendors providing these new services will be substantially different from what we know now.

The standards for this new network will no longer be established by the Bell System, although Bell will have a strong influence. The groups that set these standards will be represented from countries throughout the world, and the standards will be international, resulting in an international network.

International Organizations

There are two principal international standards organizations whose functions are to define the technical aspects of any activity that will be used by telecommunication customers throughout the world. The International Standards Organization (ISO), an agency of the United Nations, is a nontreaty, voluntary body that handles technical (or nontechnical) issues that could affect the member bodies. CCITT (International Telephone and Telegraph Consultative Committee) is the other organization. CCITT is a treaty organization whose member countries sign the convention drawn up by a plenipotentiary conference. The committee represents the postal, telegraph, and telephone authorities of the various countries. For the United States the State Department is the official representative. CCITT is a technical advisory agency of the International Telecommunication Union (ITU), founded in 1865, to serve the U.N. members.

One major example of the CCITT organization planning is the X . . . series of standards for data communication, which allows data networks to communicate with each other. An example of the ISO planning is the OSI (open systems interconnection). This is a seven-level model; the first three, dealing with the physical and link layers, are defined by CCITT.

The CCITT was founded 30 years ago as a permanent body of the ITU for the purpose of handling the various aspects of international telecommunication services. It is essentially an overseer of the various postal, telegraph, and telephone organizations within the countries it serves. It is the only official organization providing international telecommunication standards; other organizations are providing recommendations for standards.

Local Organizations

Local organizations include the American National Standards Institute (ANSI), the Institute of Electrical and Electronic Engineers (IEEE), the National Bureau of Standards (NBS), and the Electronic Industries Association (EIA).

The ANSI is a U.S. agency and a voting member of ISO. It is composed of vendor groups, research and development (R & D) organizations, standards groups and other dues-paying interested members. The various subcommittees meet every two months and their recommendations are forwarded to the full board.

The IEEE is a society of professionals from industry and academia established to exchange technical ideas and to establish standards (i.e., the current activity for LAN standards).

The IEEE-802 standard deals with the interconnection of LANs at the physical and link layers from the OSI model. The standard has several parts, each dealing with a different LAN architecture.

The NBS is an agency within the U.S. Department of Commerce and is funded by the federal government. Their communications goal is to enable various governmental agencies to pass information among themselves by independently selecting the computing or telephony equipment necessary within the agency to perform the special functions of that agency.

The EIA is a society of industrial corporations established in 1924 to develop manufacturing standards. The RS-232C interface is an example of EIA efforts.

Obviously, considerable overlap exists between these organizations because of merging technologies and common interests. In many cases the same people participate in subcommittees for different organizations. Efforts have been made to minimize the duplication. The NBS has strong ties with the other national committees within the United States. Recommendations from local committees are presented to either ISO or CCITT, who have developed liaison meetings to resolve their differences and improve their working relationships.

As required standards are completed, they are presented to the CCITT for approval. Countries and organizations can comment on these standards. The standards are developed along the layered protocol structure of the OSI. Once these standards are known, the vendors, with their inexpensive microprocessors and memories, can make these services available worldwide.

One common complaint of standards bodies is the time taken to develop any standard. The development of a standard is based on the

cooperation and consensus of the parties (countries) involved and requires a technical review of the content. This approach is fair and time
consuming and, therefore, always open to criticism.

7.2 Equal Access

The current telecommunication network, in spite of its excellent performance over the years, suffers from several limitations. Chief among
them is the requirement to rapidly introduce new technology and services as they emerge. For this reason, among others, various regulatory
telecommunication bodies around the world have decided to deregulate
the telecommunication industry. The assumption is that a deregulated
industry will respond more rapidly to the changing technology to provide better service to the customers. This new political atmosphere is
one aspect of what is required for the evolution to the new information
network. Two other criteria for this network are a clear statement of
the requirements the network is expected to meet and standards for
these requirements.

Within the United States deregulation's first challenge was to provide
access to the various long-distance carriers without degrading service.
With the deregulation of AT&T and the introduction of competition in
long-distance service, the subject of access from the local exchange to the
carrier was bound to be a controversial item. The local exchanges, for the
most part, are Bell owned and equipped, with Bell equipment making it
difficult to interface with other carriers. A substantial investment in interface hardware/software was necessary to facilitate this change. In fact, it
was necessary in many cases for the local exchange to build new interfaces to AT&T to ensure compliance with deregulation. Typically, AT&T
subcontracted the local exchange to design these links. In general, the
access from the local exchanges to the long-distance carriers is being
performed with the same high quality as the connection to AT&T.

All this was the result of the modified final judgment between the
Department of Justice and AT&T, which provides local dial access to all
interexchange carriers (IECs) with facilities and terminations that are
equal in type, quality, and price. The implementation of this agreement
was complicated by the existing structure of the network, which inter-
mingled both local and long-distance services. This effort is still going on,
requiring the network to be broken into two distinct parts: the local
exchanges known as LATAs (local access and transport areas) and IECs.

The LATA function is to provide local service to the customer and access to the IECs. Calls between LATAs must be handled by IECs.

What does all this mean? The initial subscribers to MCI, SPRINT, or other carriers were required to dial approximately 22 digits for long-distance calls, whereas the subscribers to AT&T had to dial only 10 or 11 digits for the same call. With the introduction of equal access, all subscribers are given an option to select an IEC, and long-distance calls with that carrier will require only the minimal number of digits. Access to other carriers will be via an access code plus the called number. The access code includes a three-digit IEC identification. The dialing pattern for a prescribed IEC is 1 + 10-digit number; for other IECs it is 1 + access code + 10-digit number.

The method used by the local exchange to access the various carriers can be determined by the company, but will normally include some type of tandem arrangement. The use of a tandem switch to access the carriers simplifies the amount of work involved in this access and combines traffic for the lightly loaded carriers. The basic linking arrangements between the local exchange and the carriers are shown in Figure 7.1.

Fig. 7.1. Equal Access Arrangements

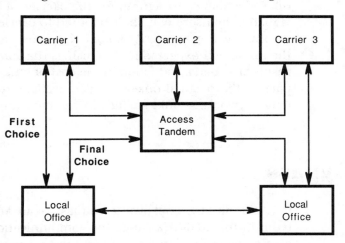

Variations of these arrangements are possible and encouraged, depending on the requirements of the local exchange and the carriers. One popular arrangement, where sufficient traffic exists, is to have direct or high-usage linking between the local exchange and the carriers with overflow to the tandem. This will minimize the linking costs. However, the access tandem is the fastest and easiest way to implement an IEC request for operation within the LATA. In many cases the tandem would require major modification, but the local exchanges can suffice with little or no change.

One of the interesting aspects of the equal access arrangement is the change it has made to the routing within the network. The long-distance network formerly operated on a hierarchical structure in which calls would go up or down routing chains according to their destination. With deregulation the hierarchical routing structure probably will disappear as local exchanges handle routing in a grid pattern, and each IEC will develop its own long-distance pattern. In addition, the following new services can now be introduced into the routing structure:

Bypass routing   The ability to go directly to the interexchange carrier without using the local exchange for long-distance calls.

Optimal routing   The use of programs to determine the optimal carrier for a call and allow for changes with time of day.

Carrier ownership   For large corporations or a business with large teletraffic volume as long as the cost can justify this arrangement.

Equal access is being currently implemented, and users are being offered a chance to sign up for the carrier of their choice. If no selection is made, the users will be distributed to carriers in the same proportions as other voting users. Although there are many long-distance carriers, the number is expected to diminish as the competition increases. The net result has been a reduction in the cost of long-distance service which was part of the original objective. The availability of new services from the new networks should surface in the next few years.

7.3 Bypass

In 1981 the FCC approved a request from Xerox for a frequency spectrum in the 10.6-GHz range for communication within a city from rooftops or a connection from a rooftop to a nearby satellite.

The ruling established what is known as digital electronic message service (Dems), which utilizes a T1 line (capable of 1.544 Mb/s) for the various applications, ranging from videoconferencing to high-speed computer communications. Actually, the FCC established two Dems networks: one for service to a minimum of 30 cities, and one for service to less than 30 cities. Cities, however, are not defined as cities but as standard metropolitan statistical areas. The Dems service to 30 cities or more is known as *extended service* and has been allocated 70 MHz for local distribution and 20 MHz for internodal links. The Dems service for less than 30 cities, known as *limited service*, has been allocated 30 MHz for local distribution and 10 MHz for internodal links.

This ruling also legalized the bypass industry. This is one example of the current activity in the bypass business. Firms are now taking advantage of this new allocation and offering video, data, or other wideband service that the telephony company is unable to respond to. Companies and entrepreneurs are looking at the bypass market as a high-growth market and a way to gather the communication required in today's environment. Many of these networks operate on the principle of demand assigned multiple access (DAMA), which means the customer will be given the bandwidth required when requested. If the customer only needs voice bandwidth, only voice bandwidth is provided; if the customer requires videoconferencing bandwidth, video bandwidth is provided.

Other items to consider are traffic mix, traffic density, and grade of service before any final decision is made on bypass.

Examples of Bypass

Although bypass has been around since the start of telecommunication, the 1981 FCC ruling and divestiture provided the impetus for tremendous growth in this area. It is safe to say that half of the bypasses have been built since 1981.

A lot of vendors, including AT&T, are in the bypass business. Some examples of their services are

- The long-distance or common carrier offers users various methods of shipping the call to the destination without using the local phone company.
- Owners of office buildings and industrial parks can offer service to their tenants by leasing lines from the phone to the long-distance carrier.
- Independent networks (for example, Dama Telecommunications), which offers nationwide data and voice service.

Some of the bypass options are illustrated in Figure 7.2.

The network or bypass is normally built to enhance the data and video services within an organization, however, it is possible that the voice traffic routed in this manner may provide the largest cost saving. Voice traffic to nonnetwork locations may route through private facilities

**Fig. 7.2. Bypass
Configurations**

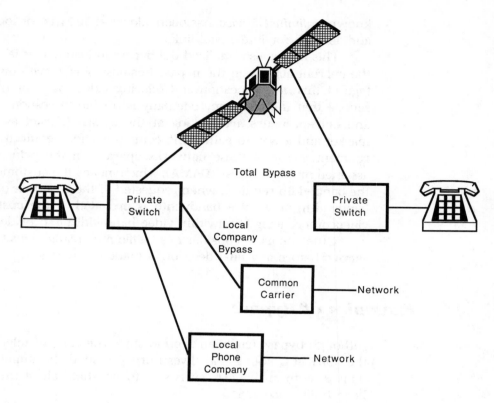

for the greatest part of the trip, with local access lines providing the final termination. The network must have the ability to route the call to an office within the network nearest to the destination and then to have the call routed off-net via available access lines.

The data portion of the network will require computer-to-computer communication on a high-speed link in addition to the requirements for the various terminals to interface with one another and the mainframe. Remote batch input and output are also required. Freeze-frame or full-motion videoconferencing is needed for various applications. It's good to be cautious in this area until the actual needs for this communication can be determined. There is nothing worse for the future of private networks than to have a load of expensive video equipment sitting around collecting dust and costing money.

Each service must be examined on a case-by-case basis for technical and economic merits before it is included in the offering. A trial site(s) is then selected to determine how well the service is received and to test the equipment.

One critical aspect of a private network is the billing methodology and capability of the system. The corporation will want the cost allocated properly to its divisions, and the equipment must be able to differentiate points of origination and destination. The system should also determine time duration and bandwidth used by the call for proper billing.

Technology and regulation will decide whether a business chooses a bypass. The introduction of an integrated services digital network (ISDN) configuration (an international standard) by the phone companies with its voice, data, videotex, and video capabilities can negate the need for bypasses. However, the advance of these services can and will be affected by state and federal regulations. Judgment on bypasses must be deferred.

7.4 Training

A great deal of money is spent on training. It is estimated that employee training each year costs approximately half the total dollars spent on higher education. Virtually every university is conducting or experimenting with some type of electronics training. Programs can be purchased or transmitted to a special center or the home, depending on the course or university presenting the information.

The benefits of teletraining, a term commonly used for this type of training, are enormous: for example, the leading expert on a subject can be made available to an audience worldwide instead of at a local campus. Corporations are taking advantage of this situation to place expertise on tape for training and for future reference. This concept is being taken one step further, as artificial intelligence is attempting to develop programs that will simulate the way an expert would solve a particular problem. This would be a tremendous aid to any large installation that can feed a program with symptoms, problems, or breakdowns, and then receive a recommended action. This may not correct the problem, but it will probably start the maintenance personnel on the right track. Most maintenance personnel appreciate an indication of where to start their investigation.

The continuous education program presently emerging in the United States will have an influence on any training plan. More and more people are attending classes, and this is an opportunity to make education available at home.

The use of schools for training will still provide an excellent way to interface with other people within the organization. My experience has been that more is learned outside the class at these special weekly training courses. The use of teletraining at home would not diminish the need for schools, but it will probably inhibit additional growth in class-room study.

The easiest way to establish a training program is to buy a video camera, recorder, and some tapes. Taping a lecture or a meeting can be a start toward a successful training program. Where this training program leads is highly dependent on the employees' reception to the program. Video is an effective way to convey information, and tests have shown that the student can learn as much from this medium as from a live instructor in the classroom. Its cost is very practical if a commercial product is used. It is highly recommended that the commercial version of video be used, because it is less expensive and the employee can use the tapes at home.

7.5 Cellular Mobile Radio

Cellular mobile radio (CMR) offers an advantage to both the user and the vendor providing the service. The current CMR allows the user to be located anywhere in the serving area, as long as phone calls can be received, without worrying about frequencies (the problem of previous systems). The vendor can enter a major area of telecommunications with a low getting-started cost. Don't confuse CMR with those wireless band-sets Jack Nicklaus is always pushing. With those, the transmission is restricted to a certain distance from the normal phone and the call to the phone is established over the pair of wires between the home and the switching office. With mobile telephone service the call uses special frequencies between the office and the unit for the call.

Another major advantage of CMR is that an individual can get into this aspect of telecommunications without a lot of capital. The vast majority of telecommunication business is extremely capital intensive due to the large investment necessary in wire and cable. It always amazed me that some of the people I met in telephony, who could easily prove diminished capacity to a murder trial jury, were in charge of $100 million budgets.

Method of Operation

Cellular mobile radio operates by dividing a city or serving area into units or cells, each cell able to transmit at certain frequencies within a prescribed range. These frequencies can be reused by other cells as long as there is sufficient distance between them to avoid interference. This principle is the same as that used by a local radio station transmitting at a certain frequency, yet that frequency could be reused by another station elsewhere in the country. The exception, in radio, to this is clear channel stations, which means no one else can use that frequency. CMR does not have clear channels yet, but when they want to transmit mobile calls across the country, I am sure they will discover this concept.

A cell serves an area as small as a mile or as large as 14–16 miles, depending on the situation. This is illustrated in Figure 7.3.

Fig. 7.3. A City with Cellular Mobile Radio

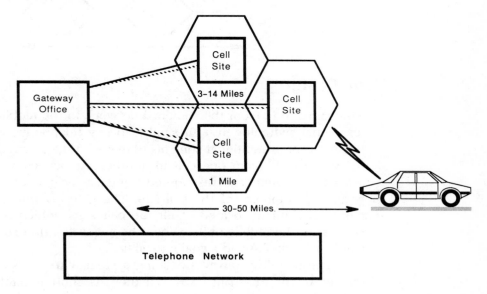

Each cell is equipped with a transmitter/receiver (a tower) for communicating with the mobile unit and telecommunication equipment for connections to a special central office. The central office will connect the call to the local phone network or to another radiotelephone. This central office is also responsible for monitoring the strength of the signal from the transmitter/receiver to the mobile unit. When the signal is unsatisfactory, the central office will pass the call to a neighboring cell capable of a strong signal. The customer normally is unaware of this transfer, which allows for constant good communication as he or she drives across the city.

The cell site arrangement is used (hence the name "cellular") to minimize the number of frequencies needed to serve an area. The cells are arranged as shown in Figure 7.4. The cell array allows the frequencies to be reused as quickly as possible without causing interference between calls. Should the number of mobile units grow within a cell, the cell can be subdivided into additional cells to increase service.

Fig. 7.4. Cellular Cell Arrangement

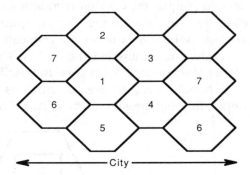

System Design

The connection from the standard network to the mobile network is an electronic switching system similar to any telephony switching system but with some additional controls to measure the strength of signals and to switch calls from one cell to another. Associated with the gateway switching system are distributed controls located at the transmitter/receiver equipment of the cell site. The gateway office signals to electronic equipment located at all the cell sites, which in turn pass the request to the mobile units. When the requested unit sees its number, it alerts the cell site and signaling starts.

The mobile gateway switching system with cell site control is the normal method used in CMR, but there is another method, which employs offices (smaller in size) at each of the cell sites. One of these offices is designated as the gateway office to interface to the telephony network. This method is attractive when the number of cell sites is small.

This CMR improves the current system through improved service, more channels, better availability of channels, and a growth scheme that should be able to keep up with the demand. The growth of the service is now tied to the cost of the unit in the automobile. Present indications are that the cost of the unit will drop and the features of the service, together with acceptance, will increase.

7.6 Teleconferencing

John Naisbitt, in his book *Megatrends*, concludes that "Teleconferencing is so rational it will never succeed." I was reading this prediction shortly after I had concluded that videoconferencing was so irrational it was destined to succeed.

I'm not sure how John Naisbitt reached his conclusion, but I keep thinking of the presenters I know who lean on the projector when giving a talk with overheads, or stand between the projector and the screen (which casts shadows and half-images everywhere until no one is listening to what is being said), or who never allow you to see any part of their body except the top of their heads as they read their talks. The thought of watching these people on videoconferencing with an audience expecting Johnny Carson or Roger Mudd is so repulsive and irrational, I concluded videoconferencing is here to stay.

Videoconferencing is in a period of what I call rhetorical exaggerations. Authors are predicting a tremendous success or a dismal failure for this service. We know videophone or Picturephone was not a success in the 1930s or in the 1970s. Why should videoconferencing succeed in the 1980s? Before any conclusion about videoconferencing is drawn, I will explain audio conferencing, sometimes referred to as teleconferencing, and full-motion videoconferencing. Freeze-frame or slow-scan video will not be explained, because they are interim steps between audio teleconferencing and full-motion video.

Audio Conferencing

In early telephony the operator would frequently set up a connection among more than two parties. However, as automatic switching was introduced, the use of audio conferencing declined as operator involvement with all connections decreased. The recent introduction of stored program controlled (SPC), or computer-controlled, machines has led a revival in audio conferencing because these machines have more capabilities for this type of service. AT&T recently announced an audiographics conferencing service with controls from their various SPC systems.

Audio conferencing is an excellent way to hold meetings or reviews when the participants are familiar with each other and multiple locations are involved. An audio conference can be as small as two groups of people gathered around speakerphones or several meeting rooms all connected. The conference can be established through a PBX, an operator, an SPC switch, or a vendor who provides this type of service. Your choice will depend largely on the type of telephone switches available at the locations. Most modern PBXs will have as a built-in option some type of audio conferencing capability.

A user-initiated conference normally is established by the originator dialing each of the participants via the PBX. Standard equipment for this type of arrangement works well when the number of participants is small. For larger groups or when people are in different locations, a "meet me" conference can be used. Here the participants dial a special number at a prearranged time. This arrangement allows a "bridge" to be inserted in the connection. A bridge eliminates feedback and distortion by ensuring that the input and output are not switched at the same time, although this special switching is not noticeable to the users. Operators can also establish these conferences.

Sophisticated conference bridges are available from telecommunication companies that specialize in this service. They also offer special conference phones for meeting rooms. A conference phone is an extension of the basic input and output capability of the telephone. Inputs are allowed from the microphones combined in a mixer, which feeds the signal to the transmitter. The system cannot handle all the mixers at once (the load is too great); a push-to-talk switch is used or a microphone is turned on when someone starts to speak. The latter method is preferred, but it is more expensive.

Enhancements to the teleconference include electronic blackboard and graphics. Potential users should investigate these enhancements when considering audio conferencing services. The main advantage of the audio conference is its ability to use the existing telecommunication network. By promoting this service properly, travel savings can be realized without a large investment. Most people are unaware of audio conferences and their capabilities.

Videoconferencing

For structured presentations where the presenter is attempting to sell an idea or a product to management, a customer, or a group of people unfamiliar with the idea or product, the audio teleconference only conveys part of the message. In addition, it inhibits the question-and-answer part of the presentation. For this type of conferencing full-motion video should be investigated.

The introduction of satellite communication has been the biggest enhancement to videoconferencing. A satellite provides the necessary bandwidths to transmit video. An audio bandwidth is 3000 Hz, whereas a video bandwidth is 6,000,000 Hz. The present analog phone network cannot transport such a large bandwidth. However, technological advancements in telecommunications are changing audio and video to a digital transport. The audio requirements in a digital format are 64,000 b/s; the video requirements are 1,544,000 b/s. This is compressed video, but it does provide full motion and is consistent with the basic building blocks of the digital network.

A videoconference connection is set up as in Figure 7.5.

The typical reason for justifying videoconferencing is the reduction in travel. Personal productivity is an even better justification for the system. Productivity increases as a result of videoconferencing is difficult to measure in dollars and cents, but a few examples might illustrate the point. Corporations are headed by a few decision-making executives whose time is too valuable to be wasted sitting on airplanes for hours to attend a one-hour meeting.

The second example relates to the major development projects going on in any company. The system development in most cases is spread out across the country or across the world. Holding system reviews is difficult and time consuming, and the number of people that can be sent to one of these reviews is limited. Hence the reviews proceed through a superficial agenda.

Videoconferencing can promote more frequent system reviews, make reviews more meaningful, and encourage better participation. Considering the tremendous cost of major projects and any delays, a videoconference system could pay for itself quickly.

A third example is the hard-to-get speaker or the executive who is trying to develop or maintain a "culture" in a large organization. An effective way of dealing with the infrastructure of an organization is via videoconferencing because many people can be reached quickly and information can be conveyed accurately.

The final example concerns the ability of videoconferencing to present information to a customer without the customer traveling to the vendor's facility. Consider the possibilities for the field salesperson when the presentation can include a question-and-answer session with the technical guru from R & D together with tapes showing the system in operation, in manufacturing, or in both. The competitive advantage that this offers should generate great interest in special videoconferencing arrangements.

Many people will argue that the videoconference cannot replace going out "slapping flesh" or observing body language, and I would agree with them. Videoconferencing will never be a replacement for face-to-face communication. Like most electronic aids it is a tool to supplement such meetings and to improve the productivity of the salesperson. For example, a meeting between an important customer and an executive presently takes about two weeks to set up; with videoconferencing it can be accomplished in a couple of days.

Fig. 7.5. Video-conferencing

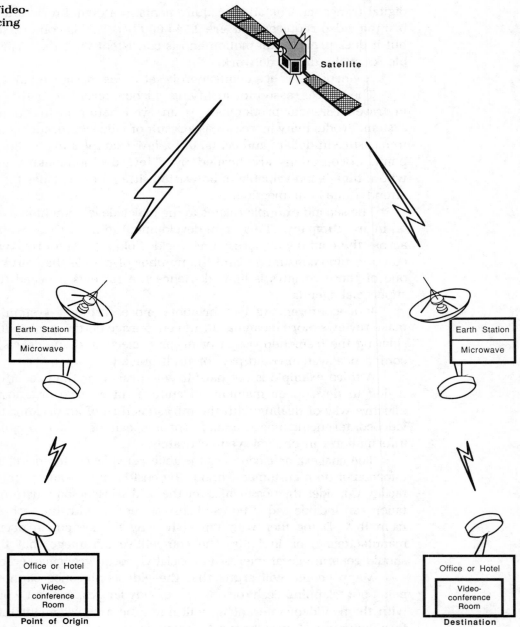

How It Works

The transmission techniques for videoconferencing can be analog or digital. The analog technique is used to transport TV to homes. It requires a wide bandwidth, which is expensive and available only to large commercial enterprises. TV transmission time, however, can be leased from the telephone company or from satellite corporations if it is available, but this approach is expensive. There is also the added cost of getting the signal from the company to the telephone company or satellite.

A better approach is to consider the digital technique that samples the analog signal with a special codec (a device for coding and decoding the analog signal) at an acceptable rate for transmission toward the destination. The codec will normally sample and transmit the signal at 1.54 Mb/s, which is compressed when compared to the analog signal but acceptable for full-motion videoconferencing. The 1.54-Mb/s rate is what is known as the DS1 rate, and numerous facilities are available to cater to this rate.

The digital network and most digital switches are built around the DS1 rate, and voice conversations that are converted to digital transmit at 64 kb/s, which is approximately 1/24 of the 1.54-Mb/s rate.

The codec used to convert the voice to 64 kb/s is based on a telephony standard, which means two parties can communicate although they are connected to codecs from different manufacturers. This is not the case with video. Most vendors' codecs are incompatible with the compression techniques of other vendors. This incompatibility will inhibit the growth of videoconferencing until standards are established, agreed upon, and adopted by domestic and international organizations.

The basic technical problem is that a standard of 1.54 Mb/s was established, but new manufacturers were able to obtain and transmit a better-quality picture at the same rate or employ a lower rate with the same quality. This is a never-ending dilemma, because technology can always develop a slightly faster or better scan method. The solution is to establish standards and allow the manufacturers to build codecs to those standards. Some may argue that this would inhibit technology, but a transmission rate of 64 kb/s for voice conversations has not inhibited any activities in voice communication. The problem is further complicated because Bell can't in the present environment set a standard and expect anyone outside of Bell to follow it.

Implementation

Implementing the videoconferencing system is not easy. Subcontracting parts of the system might be the best approach. There are four major areas to videoconferencing: the room, the video equipment, the network access method, and "selling" it to the organizations.

The videoconference room is normally set aside for videoconferences although it can and should be used for teleconferencing. It is equipped with a tabletop control panel. At least two monitors should be provided: one to show the room, and one to show the remote site.

The video equipment is relatively standard and can be purchased from vendors. Mixing vendors could create compatibility problems, however, because vendors don't use the same codecs.

Network access provisions depend heavily on whether the network is private. A connection to the phone company, a "bypass" microwave to a long-distance carrier, or a disk at each site connected to a satellite are all feasible and should be evaluated.

Selling videoconferencing to the corporation may be the most difficult part to implement, because a great deal of training is necessary for people using the equipment and for people making the presentation. With regard to the people making the presentation, it appears as if the world is divided into two groups: those who freeze in front of a camera, and those who act like Howdy Doody. Both need training and practice. No one should make a presentation on video without preparation and coaching. Past experience has shown that the self-conscious attitude most people have with video did wear off and an enhanced feeling of proximity and intimacy with the other parties was eventually achieved. There is no substitute for practice and training for getting people to feel comfortable about a new way of communicating.

Videoconferencing can be economically proven, and the technology makes it extremely feasible. However, its success will depend on whether the customer views this service as a broadcast service or as a person-to-person service. Broadcast services are TV, radio, and newspapers; and person-to-person services are telephony and mail.

From a customer standpoint these services are distinct. One reason for the slow growth of videoconferencing is the public's attempt to reconcile which type of service videoconferencing is. (Many early telephony installations in farm communities were started in order to keep track of conditions affecting farming, until it was discovered that radio was the preferred method for obtaining this information.) Which type of service is videoconferencing? Both. Will it succeed as both? Probably —but to a greater extent as a person-to-person service than as a broadcast service. As a broadcast service, its most popular outlet will be special rooms in hotels and motels where presentations will be made to customers, speeches will be made by politicians, operations by leading surgeons will be viewed, and so forth. In other words, the customer will be expecting a quality show when he or she travels to a much ballyhooed presentation.

Person-to-person communication will be much harder to sell and will have a slower growth rate. It may take the better part of a generation for people to accept videoconferencing as a genuine method of communication. But videoconferencing is here to stay, and the sooner it is properly recognized the better. Eventually, it will be more successful when used as part of a private communication network. Five things assist this growth:

1. More private communication systems will be installed.
2. More 1.54-Mb/s channels will be available.
3. More inexpensive earth stations will be available.
4. The ISDN will be established.
5. Greater use will be made of fiber optics as a transmission medium.

What then of the presenters exemplified by the mumbling speaker with his head down? We will either retrain this type of presenter, or we'll accept them at a distance the same way we accept them today when they make presentations—we will try to listen to the material and ignore the style.

7.7 Storage Systems

Ever since the earliest day of telecommunication, the connections through the network have consisted of a transmitting party or terminal and a receiving party or terminal. A variety of storage systems is appearing in the business that will allow the message to be transmitted and the party to receive it later. The impetus for these services arises from the current frustration with mail delivery and playing "telephone tag" with someone over several days. The most popular of these are voice mail and electronic mail. This is not to ignore store-and-forward systems, which have been around for years.

Voice Mail

On large PBXs or networks it is more economical to provide voice mail for the system rather than individual answering devices heard in many homes these days. Within a network it is advantageous to have a central-ized voice-mail system or systems for the various PBXs from one vendor. Voice-mail systems have a variety of features and methods of operation, and the user shouldn't be required to learn each voice system.

Calls can be forwarded automatically, or upon busy or no answer, to the voice-mail system, which will provide an introductory announcement (generic or personal or both) to the caller, ending with a request to present the message at the tone. The user of the voice-mail system (called party) can be periodically called, have a display that a message is waiting, or call the unit to determine whether there are any messages. The latter requires the use of an identification code. The approach is based on the system employed.

Voice mail is a viable service in current telecommunication. Realiz-ing this, vendors are providing more and more features with each re-lease. Some services worth investigating are as follows:

- The PBX should be capable of distinguishing between call forwarding and transfer to voice mail.
- Voice mail should be capable of transmitting an indication (light) to the party that a message is waiting.
- Timing on the system should be adjustable by the user.
- The system should provide transfer to a secretary at certain times and to voice mail at other times.
- The voice-mail system must be capable of interfacing to different PBXs.

Over half of the calls within a PBX are not answered by the re-quested party, because the party is busy or not at the desk. It is also known that a large amount of traffic is not generated because "no one is there at this time." A voice-mail system is the answer to these frustra-tions. Busy executives can use this service to dictate letters to their secretaries while traveling or working late.

Electronic Mail

Electronic mail covers numerous services that are available or that are becoming available. Electronic mail is best defined as a telecommunica-tion service that allows its users to exchange messages and documents electronically with speed and accuracy between office text machines (electronic typewriters, word processors, and computers equipped for transmitting and receiving documents and messages). Electronic mail is voice mail from a terminal.

Approximately 70% of mail is generated by computer, so time and money can be saved if this information can be transported electronically. The service offers great promise for office and home. It is besieged, however, with standards problems, which will slow the growth anticipated for this interchange of information. The lack of standards is forcing many private networks to adopt their own standards, which will make the interfacing more difficult in the future.

Internationally, the service is referred to as *teletex* and is an extension of telex service. Telex service is probably one of the oldest services that allows communication between two parties without voice, that is, from one teletype terminal to another. Telex has had international standards associated with it, and it is important that similar standards be available for teletex.

Fig. 7.6. Local Communication Scheme

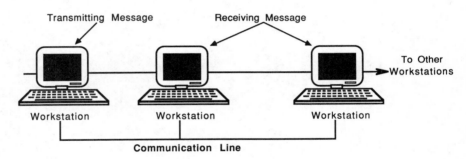

In Figure 7.6 the terminal transmitting the message merely types the message and indicates who will receive it. Shown are two terminals located in the same area. If the terminals were located at a distance, modems would be used for transmittal over the voice network, or a data network would be used without modems. The party away from the desk should receive an indication that a message is waiting.

7.8 Facsimile Service

Facsimile service (facs) is the transmission and reproduction of printed material or graphics at the distant terminal. Reproduction is a necessary part of the definition (I have sent documents via facs that are still not reproduced). This service has been available for years and is floundering due to a lack of standards. A consistent problem of facs was determining whether the machines within a company were compatible, and this problem was compounded when transmitting documents between companies.

CCITT Group 3 is working on international standards for facsimile, and I am sure this will improve the situation. It will be necessary to establish a national standard as well, a subset of the international standard, that vendors can use.

The future of facsimile with ISDN will probably be as part of electronic mail as a 64-kb/s channel becomes available for this service.

8

Current Network Environment

Having stood on the vestibule of networks for a considerable time, we are ready to enter the mansion itself. Networks can be viewed in terms of cost, service, traffic, or routing, among other things. A clue as to what this mansion is was given earlier when we spoke of the user's needs. It behooves us to ask what problems the user will have in order to understand the approach to the network's solution. If no data traffic exists, and there are no plans for this traffic, then a voice/data configuration is not the answer to the problem. Let us begin with an awareness of the needs and then progress to the services and features.

8.1 Network Needs

Appropriately, many of the requirements for a network are determined by the organizational structure and its functions. Organizations can be divided into many categories, but for purposes of telecommunication criteria we use large infrastructures, strategy oriented, and state-of-the-art.

Large infrastructure organizations probably dominate businesses throughout the country, handling a wide variety of products. Infrastructure groups usually are centralized companies that have a voice network used by everyone for on-net calls (calls within the organization). In many cases these calls are through an operator. Off-net calls are handled via an access code from the PBXs. Incoming calls from off-net points typically are handled by an attendant unless centrex or direct-in-dialing is featured.

For data the company encourages everyone to use IBM. A large data processing center can be reached through a systems network architecture (SNA) arrangement, although there are local capabilities.

The problem with this arrangement is that both the telephone and data processing systems have been slow in recognizing the importance of local area networks (LANs), data networks, and voice options. Thus most new data configurations have moved clandestinely into sections of the organization to solve a particular problem without any awareness of networking or interfacing with other units.

However, communication must be totally solved, not divided into local and wide area services. This division is performed without regard to voice or data. Because approximately 60–80% of an individual's information is obtained from local sources, good local voice/data communication within these organizations is necessary. However, good long-distance voice and data services are still required: this necessity has not been diminished by this statistic.

A large infrastructure organization also is blessed (?) with a voice communication group in addition to data processing or information management. Both groups argue that only their organization is capable of understanding and handling the telecommunication needs of the company. Typically, the infrastructure organization needs technical and political solutions to the problem. The technical solution is normally in the form of a configuration suited to the business and approved by higher management. The implementers will have the political problem to overcome.

Strategy-oriented groups (banks are a prime example) would like a network configuration giving them a competitive advantage over other savings institutions. A sufficient risk is involved in this, because the network is, almost by definition, unique, and performance standards are not available. Should the network fail, the problems are enormous. Data is more important then voice because the updates of stock information or money transactions are the bloodline of these networks. For years these networks have been private because vendors were hard pressed to build such unique applications. Recently, the use of common transport for these networks has become more evident, and this trend should continue as wideband fiber-optic networks are installed.

State-of-the-art groups do not have much time for studying and understanding telecommunication. They are only interested in reducing the cost of communications while improving productivity. To vendors they are probably the largest group and include most universities and small-to-medium businesses for which communication is a support function, though a critical one. This group presents the most volatile category, because they are subject to rapid changes in budget and administration. The group also depends heavily on vendors for the latest system information and for analysis of their needs.

These three categories complicate the effort of any vendor to address the total telecommunication market with a single switch or family of switches. They also illustrate the diverse aspects of the market from a traffic administration viewpoint. A major task is to understand and administer the interfaces between the products serving these markets.

8.2 Determining Requirements

Within the large infrastructure organization the current business and technological environment dictates that the products for the network have every gadget and feature ever conceived. The system manager must decide which features are necessary for the operation and deal with the skepticism existing between the telephone and information processing personnel. Both groups will require special features in the network, and the information processing group will be reluctant to participate in an endeavor that may jeopardize a kingdom. This is a very real situation and must be dealt with. In addition, some consultant has convinced top management that money can be saved by combining data processing and communication systems. Consequently, top management can't understand why the network handling voice and data isn't in place.

One answer to this dilemma is to have a representative from data processing on the evaluation team and arrange to have one of their favorite programs passed through the network at a greater speed. There is nothing like speed and more speed to impress information processing personnel. Many times I have sat through a lunch listening to a boring description on how a new load will increase the response time 0.05 ms. If the favorite program transmits at 1200 b/s, make arrangements to transmit it at 9600 b/s and watch the response.

Presently, one victim of this conflict is AT&T. AT&T has very fine processors, but they are having problems selling to information processing personnel for two reasons: (1) they're not IBM; and (2) they're from the phone company. I am sure AT&T will overcome these problems, but it will require time and a proven track record. It is hard to imagine a 100-year-old company as big as AT&T having to prove itself. Some of AT&T's problems can be traced to their zeal in building a strong voice network and making data subservient to voice on the network. It appears that this attitude has changed.

The current "voice" communication systems are shown in Table 8.1 and cover the various applications that have evolved.

Table 8.1. Telecommunication Options	Service	Function
	Key Telephone Systems	Key-selected access at a customer station to a multiplicity of links, offices, or other lines. Hold, intercom, and other features available.
	PBX Service	Station-to-station calls without public network. Attendant or automatic in-dial for incoming calls. Outgoing via access code. Some systems have automatic identification of outgoing by station.
	Centrex	Station-to-station calls with automatic access to/from public network, attendant service, and out-dialing identification.
	Central Office	Customer-to-customer, link-to-customer, and customer-to-link calls are handled. May provide operator service and PBX access. Can also provide access to common carrier for long-distance calls. Billing service included. Many operate with remotes.
	Tandem Office	Provide link-to-link connections. Also, access point for common carriers. Relay point for digital synchronization.
	Toll Office	Combine functions of central office and tandem.
	Traffic Service Position System	Operator service for local and long-distance calls. Automatic billing on long distance.
	Information Service	Access to local or remote operator for calls.
	Automatic Call Distribution	Incoming calls are uniformly distributed over answering positions.

However, data and the incorporation of data into a network is more recent and require a closer analysis. Three major developments in the past 15 years have focused on moving data in a uniform structure through a network. The first change was the packet-switching network, introduced to move data through a network in a format substantially different from voice. These networks are rapidly growing as a viable means of passing low-speed data from point to point.

The second change was the 1974 introduction of IBM's SNA. This presented a standard not only for the IBM systems but for any machine interfacing to an IBM mainframe. It was a major advance in moving data around a corporation or university in an orderly manner.

The third major data change was the introduction of Ethernet by Xerox, which affected the way data is moved in a local environment, and, with it, the start of LANs. The LANs have seen phenomenal growth but only token effort to establish standards. Standards are vital if these units are to move into other environments.

These three "revolutions" have, for the most part, been mutually exclusive of each other with little impact on the voice network. The PBX, for example, has been moving from an analog, regulated environment into a digital, unregulated arena virtually without regard to what has been happening at the end-use level. The PBX must now "learn" to cater to all of data needs just when the standards have been established for a feature-laden digital circuit-switched voice machine.

One other change is occurring—the introduction of workstations into the market. These appear to be divided into either a data terminal with phone capability or a phone terminal with some data capability. This workstation will require access to a voice/data switch, period. The present PBX has a voice-only analog interface to the line, which converts the analog to digital at the switch side of the connection. The workstation will, if properly handled, have one single digital interface controlling voice and data.

8.3 Integration of Voice and Data

The best place to start an evaluation of the network requirements is at the customer's interface point—the PBX or LAN.

The most widely discussed topic in the PBX/LAN marketplace is voice/data integration and how it will occur. There is no question that it will occur; we don't know how, when, or by whom.

Let's look at an example.

A company has decided to centralize its computer functions and consolidate the facilities among several locations, which now consist of data and voice links. Most of the users have data terminals and phones at their desks, but the company would also like to eventually introduce workstations to combine these functions.

Before workstations evolved, the main communicating devices in the office were the telephone and the mail. Mail communication is rapidly changing to information via the workstation/computer over a separate network. The PBX is having a mailbox unit attached to allow certain users to have their unanswered calls routed to this device for voice storage.

Voice switching requires a circuit-switched facility at 4 kHz, whereas low-speed data can operate effectively on a packet-switching facility and high-speed data needs substantial bandwidth beyond 4 kHz.

A data network can effectively transmit a letter but would have problems with guaranteeing a voice connection. The LAN would have a problem with the letter if sent outside its domain and with the local voice connection.

The present generation of digital PBXs offers some type of voice/ data integration typically with a special interface and on a circuit-switched basis. These arrangements permit up to a 56-kb/s bandwidth, more than enough for everyone but the most demanding user. This is an expensive solution, however, and if the data users only require occasional connections, a data multiplexer for those terminals is a more viable solution. The multiplexer will probably require the installation of a coaxial network for the user, but, thanks to cable TV, coax is relatively inexpensive and the principal cost will be the labor associated with the installation.

Whoever pursues the history of a single switch for all traffic will find processors and associated equipment imbedded within the system rather than one all-knowing computer. These processors will, most likely, independently handle voice (circuit switching), low-speed data (packet switching), and LAN interface (Ethernet or ring compatible) with auxiliary access to a mail facility. In their general character, systems will be perceived as universal by the users.

In addition to the problems associated with the handling of these techniques, the switch must have a comprehensive management system, which is now a sine qua non part of the business. At this level the system also must be viewed as universal, permitting commands for any traffic or control to be placed from one terminal. If the switch reaches the level of importance described here, the tolerance for error will be extremely low both locally and end to end.

8.4 PBX Concepts

The PBX can be implemented in several ways, and it is not clear whether one vendor will solve the network configuration problem or whether there will be a combination of switches. The incorporation of data features, for example, into a simple voice PBX appears to be a herculean task, but switches are now being developed to achieve this result with a corresponding high cost.

The transmission of voice information is an absolute necessity, but voice can be sent in analog or digital format, although the latter is highly recommended. The call duration is normally about 3 minutes but can vary, depending on the application. Delays associated with the voice should be kept under 250 ms, the equivalent of one satellite hop.

Voice mail should have the same characteristics as the voice conversation although the duration will be less—approximately 30 s. Nonvoice mail can be transmitted via a packet or circuit network, whereas the voice-mail connection will, most likely, use the circuit network because it started as a voice call.

The entry, editing, and printing of documents, commonly referred to as word processing, are normally handled via a LAN. Occasionally, the transmission of these documents is required, in which case, either a circuit-switched or packet network is viable. The typical length is approximately a page and a half, although it can vary substantially. The interface between the LAN and the communication network is the key consideration in the transportation of word processing documents.

Another important interface with the communication network is with the mainframe processor for transferring voluminous record files between locations or running programs between these locations. The bandwidth is important here, because there is a direct relationship between the bandwidth transmission speed. The use of the facility for this transfer will probably be divided between the data information passed in a noncritical time frame (e.g., bulk data late at night), and critical transfer of information when programmers or application people are using the equipment. The transmission rate for the latter should be at least 56 kb/s.

The most critical service from the standpoint of bandwidth is the teleconferencing or videoconferencing requirements. The bandwidth for video is at least 56 or 64 kb/s, depending on the equipment employed. The duration of these conferences is typically 10 minutes. This is a very expensive service to provide and one that can be subcontracted if the requirement for it is minimal. Freeze-frame is a poor alternative.

The resolutions for the routing, voice, and data communication situations probably are sufficiently time consuming, and teleconferencing can be deferred.

8.5 Line Features

The telephone was at one time similar to the Ford Motor Company: that is, you could have any color phone you wanted as long as it was black. Even after color was introduced, service on the line was the same for everyone until "touch-calling" arrived about 20 years ago. A few additional features have been introduced since then, but no dramatic change has occurred in line service. The argument against change has always been "the installed base can't handle it."

While Bell has been sitting on its installed base, customers are wondering when the various features of the information age are going to be available. One problem is the inability of the local loops to handle a transmission rate of 160 kb/s, which is necessary for ISDN (integrated services digital network) services. Rather than adopting the 160-kb/s rate and attempting to solve the loop problem, a committee has been formed to look at various coding schemes. I think a better answer is the ISDN standard and remote units or fiber optics for customers requesting special service beyond the working limits of the 160-kb/s rate.

The business customer presents an even larger problem for the telephone company: As demands for services and features are fulfilled on PBXs or centrexes, the same facilities will be requested in the home. Many homes are at the far end of the local loops, which offers the greatest challenge to the industry. A popular feature of a PBX and centrex is the off-premise extension that allows the business executive to communicate with the company from home. If a company is equipped with electronic mail, there is no better way for a much-traveled executive to keep up with work than to install a terminal at home. A phone and modem may be the answer now, but it won't be long before requests for workstations appear.

Other line features will continue to grow as services expand. Direct in/out dialing (DIOD) is a growth rather than an enhanced service. The same is true for FX (foreign exchange) and bypass lines. These services have been around on a limited basis, but increased demand for them may tax some systems.

8.6 Planning Voice and Data Communication Systems

An average user of data is equipped with modems to transmit low-speed data over the telephony network and with multiplexers for linking data facilities to the mainframe. Everything is working properly, but people have been telling the user that networking is the way to go. That's fine, but no one is telling the user how to incorporate present equipment into a network, much less how to build a network configuration around the requirements.

Those people saying that an integrated network is the only solution are selling pipe dreams. The business network right now should be conceived as supporting three networks: voice/data, information management (IM), and monitoring and controls. The latter may be the area where the greatest savings can be realized, although little is being done. Most modern digital systems are equipped with, or provide, an option for a test access to lines and links; most computers will provide an indication of the performance of the machine. However, the operation and maintenance of these machines are supported by separate centers and individuals. These costs can be greatly reduced if the centers and individuals are combined.

The IM portion of the network is separate and can remain that way until fiber optics, with its almost unlimited bandwidth, is readily available. The IM network is probably equipped with control mechanisms and built around IBM or some other vendor architecture. To attempt to incorporate those facilities into other requirements is a difficult task; even if it could be planned, could the company afford the price and time? Arrangements soon will be available to incorporate this aspect of the network into one service unit, but that is after several other steps.

In planning new data communication facilities we should assume that at some future date the networks will merge and thus plan the facilities accordingly. However, the first priority is to establish and control the voice/data network. By data we mean low-speed terminal-to-terminal or terminal-to-computer data, which need not include computer to computer.

Today the word "digital" is critical to telecommunication planning, although it is an oversimplification of the situation even as "plastics" was 20 years ago. Digital is a great way to transmit any information through a network. However, if certain analog facilities are presently providing good service at a reasonable cost, there is no good economic reason to replace them until there is a definite need.

Integrating data functions within a business is known as *office automation*. It is a field in which everyone who makes digital equipment (LANs, word processors, or telecommunication) indicates capability. When considering office automation, we can approach the subject in many ways, but the best place to start is the terminal or the user. This area is where we want increased efficiency, improved services, and less dependency on paper. Three types of terminal equipment with voice capabilities are available:

1. A standard telephone instrument
2. A personal computer or multifunction terminal
3. A workstation that combines types 1 and 2 in the same unit

There are also interface requirements for the system. They are a single pair of wires from the terminal to the switching equipment and 64-kb/s connection capability. The data and voice connections are shown in Figure 8.1.

Fig. 8.1. Data/Voice Connection

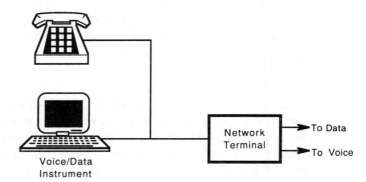

Voice/Data
Instrument

The terminals will have local and remote requirements, but we consider only local characteristics. The terminals should have access to voice messaging, work processing, facsimile, computer programs, calendars, and other aids, depending on the needs of the offices. It is still assumed that a TSO (time-sharing option) terminal is equipped in the office and tied directly, or through a concentrator, to the mainframe computer. These terminals will continue to be used by individuals who require full-time access to the mainframe. The other terminals can access this computer via a PBX or LAN on a slow-speed basis. By differentiating these needs, we will be able to expand the network in gradual steps without disrupting the environment or proposing astronomical figures for replacement. The latter consideration may be the most important thing to keep in mind.

8.7 Types of Networks

The network designer needs to know what type of network has to be designed and how critical the network is to survival. For example, the network is more vital in stockbroking than in software production.

Unfortunately, it is not possible to select a network the same way you pick a car. PBXs, central office switches, data switches, or LANs have not become the consumer's commodity the way PCs did around 1980. There are strong indications, however, that this situation may change. A few years from now, businessland may be able to install a network as easily as a LAN.

Data Networks

Data networks are available to fulfill almost every data need a company can have. If the business has large computers distributed across a state or country and wants transparent access from the users, then an IBM SNA network is probably the answer, especially if most of the machines are by IBM. This network allows the user to operate on the processors without knowing whether the work is being performed on a local processor or one located across the country. There are variations of this network and the access methods to it, because everyone will not be located with one of the mainframe computers.

A second data processing network is the LAN, an excellent answer to the problem of moving data around a building, a campus, or any area where the users are close to each other. Several LANs are available and are worth investigating.

Data networks are available to switch data between points or between mainframes. These networks are especially useful where access to programs is only occasionally necessary. Access through the mainframe in this case is not economically practical. Data networks normally route over facilities separate from those used for voice or computer to computer transmission. Data networks operate on a packet-switching principle, and the user only pays for actual usage; in the circuit-switched voice network, usage is counted whether or not information is transported. Most LANs also operate on the packet-switching principle. Where local area and data networks can be interconnected, the terminals within an office are receiving excellent service. Although this is a practical solution to the expansion of the terminal horizon, the push today is to combine the PBX and LAN. There are advantages to this solution, but the local area and data network possibilities should not be forgotten.

Voice Networks

A voice network is a voice network is a voice network, and there is nothing else to worry about. Not so! More options are available today in voice networks than Horatio ever dreamed of. The cost per circuit-minute for the typical long-distance call was $0.60 to $1.00, but since deregulation it has been dropping steadily via bypass, WATS reselling, private network, or common carriers. Bypass, for example, provides several options depending on the customer's needs. Some are shown in Figure 8.2.

Fig. 8.2. Various Ways to Bypass

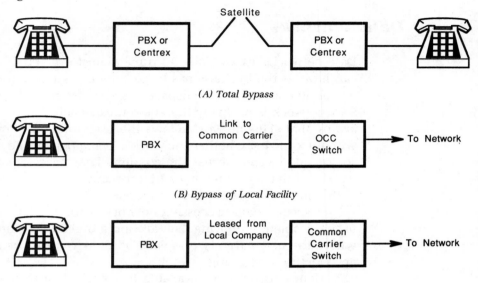

(A) Total Bypass

(B) Bypass of Local Facility

(C) Bypass of Local Service

  The bypass can range from a total bypass, as shown in Figure 8.2A where the PBXs are interconnected by satellite or similar facility, to leasing a facility from the local phone company to interface to a common carrier, as shown in Figure 8.2C. The local phone company can be bypassed by employing a link to the common carrier [Figure 8.2B]. Additional combinations are possible. The two objectives for bypass are (1) lower the cost per circuit-minute or (2) provide a service not available from the local phone company or common carrier. The main concern with bypass is its status when ISDN is offering all services at a reasonable price. This translates to a quick payback on a bypass arrangement if it is to be seriously considered.

8.8 Network Functions

User's Functions

  The user's functions in a network relate to the digital access and the ability to handle a digital telephone, a computer access, and a workstation combining the features of both. The services from a subscriber line can be abbreviated dialing, hot-line service (no digits dialed), or any number of exotic features needed for business. The main criteria today are the desire to transmit 56 or 64 kb/s to and from the line and, once this is accomplished, to introduce additional features and services.

Switching and Transmission Functions

The ability to provide optimal connections for network calls is the primary function of switching, although a billing function is needed to maintain control of network costs. The switching within a switch will usually be nonblocking and not a problem. The main question of a switch is whether the equipment has the ability to handle both voice and data calls efficiently. Switching across the network is a matter of cost, although the glut of carriers and facilities make this aspect of networking very attractive.

Digital switching and digital transmission has produced a merger of switching and transmission. A brief example on how this demarcation line has blurred recently might be helpful in understanding why switching vendors and transmission vendors are seeking the same market. Figure 8.3 shows this evolution.

Fig. 8.3. Switch and Facility Interface

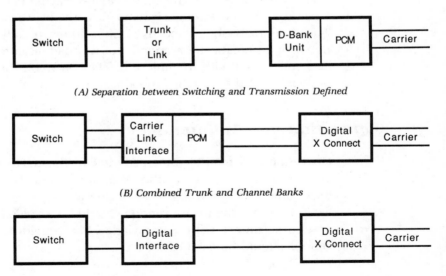

(A) Separation between Switching and Transmission Defined

(B) Combined Trunk and Channel Banks

(C) Digital Switch Interface to Carrier

When the digital carrier was introduced, as illustrated in Figure 8.3A, the separation between switching and transmission was clearly defined at the trunk (link) interface to the switch. The D-bank unit was a channel bank for converting analog to digital and had some link functions associated with it.

The transmission vendors realized they could provide their customers with some economies and gain a new market for themselves by offering units that would combine the trunk (link) and the channel banks as shown in Figure 8.3B.

The digital tandem office and the digital PBX have changed this approach, because the office is now switching calls at a digital rate and the interface to the carrier now moves to the switching side. A standard arrangement for the digital switch interface to the carrier is shown in Figure 8.3C. The digital switch has combined the switch interface function with the channel bank function, thereby simplifying the crossconnect scheme and reducing the cost.

This interface is the predecessor of the digital network of the future when many of these connections will disappear as the functions between switching and transmission blur.

Traffic Control Functions

The control of routes and their traffic-carrying capabilities must be provided by the switching systems directly or through external access. These controls should be dynamic, either initially or planned, so the network automatically responds to overload situations. The controls are also concerned with the link interfaces between the switches within the network and the switches interfaced with the network.

Operation and Maintenance

Operation and maintenance are vital to the network's well-being. They are covered in detail in other sections. These functions are especially important during the first year of operation, because problems are most likely to occur then. The vendor should be involved in the maintenance, or else a contact with a knowledgeable individual associated with the vendor should be established.

Routing

Another problem facing the modern PBX is the sophistication of the routing scheme needed to take advantage of the complex and ever-changing cost structure for traffic rates. Deregulation has spawned a tremendous increase in long-distance competition. A modern PBX must be able to take advantage of this new environment. The feature must be closely examined by anyone considering buying a PBX because it alone can be worth the price of the switch. It is difficult to evaluate a PBX on price if the alternate routing capabilities are not carefully weighed.

In addition, programs are available to allow the switch to keep up with the latest changes in the routing structure and cost. Access to these programs is free or very inexpensive. The key to this, however, is the ability of the business to change its switch when a change in the routing structure offers an advantage. Implementing this change should be achieved by maintenance or administrative personnel without contacting the original vendor.

Another essential feature of a routing scheme within a network is its ability to perform special routing under abnormal traffic conditions. These conditions can be the results of high traffic or the loss of other routes. The preferred response to these conditions is a dynamic one, where the network "reconfigures" itself to handle the traffic in the most expeditious manner. This is also an essential feature and is associated with the machine's ability to operate with CCIS or equivalent routing strategy. If reconfiguration is not provided initially, the network and the switches must be capable of adding this function later.

8.9 Examples of Network Communication

The types of connections and calls heavily influence network selection. Connections can run from the desk 20 feet away to halfway around the world; the call types can be voice, data, or video. Variations on the call types also are possible because data can include various speeds or facsimile and video can also take on several forms. However, to make things easier, Table 8.2 only deals with voice, data, and their combination. The connections run from within the building to within the world.

These combinations are increasing every day, and items not listed in the table may be capable of handling the call types and connections, but the table should serve as a guideline to network or telecommunication service. Another important consideration is whether video must be included in these configurations. Normally it is only required intercity, and special arrangements or networks can be used.

In addition to the normal customer lines and the links to other offices, a switching system is expected to handle special lines that include

Table 8.2.
Various Network Arrangements

| Connection | Need | COMMUNICATIONS REQUIREMENT OF AUTOMATED OFFICE | | |
		Voice	Data	Voice/Data
Intra-Office Bldg	Media	Wire	Wire	Wire
	Signaling	A or D	Digital	Digital
	Configuration	Star	Bus/Ring	Star/Bus
	Switching	Circuit	Packet	Circuit packet
	Equipment	PBX	LAN	PBX/LAN
Interbldg Intracity	Media	wi/fo/ra	wi/fo/ra	wi/fo/ra
	Signaling	A or D	Digital	Digital
	Configuration	Star	Bus/Ring	Star/Bus
	Switching	Circuit	Packet	Circuit/Burst
	Equipment	Digital EO	Packet switch	Burst switch
Intercity	Media	Wire/Cable/F.O./Radio/Satellite		
	Signaling	Digital		
	Configuration	Star		
	Switching	Circuit	Packet	Circuit/Packet
	Equipment	Digital	EO plus	Tandem

(a) wire/fiber/radio/cable
F.O. = fiber optics
EO = End Office

DIOD, OPX (off-premise extensions), FX, and bypass lines. The DIOD lines have been unique to centrex offices but are finding their way into the PBX market. A DIOD line has a number that is part of the public network and the private network, and this allows the customer to dial anywhere without using access codes (normally a 9 is dialed to go outside the private network). The line also can be reached by anyone on the public network.

8.10 Features of PBX, LAN, and Centrex

Digital PBX or Centrex

Several systems are competing for the office automation market, but most experts have their money on some arrangement employing the PBX in the final solution. LANs have enjoyed great success but are under great pressure to expand their horizons and provide connections to other devices. The question remains as to whether the PBX and the LAN will be combined to implement the solution or whether each will be a separate piece of the solution. If the latter is true, then office automation will never reach the height predicted for it. Thus, we will base our discussion on the optimistic assumption that the PBX and the LAN will eventually unite. Remarks for the PBX also apply to the centrex, but PBX is the universal term.

Many existing PBXs are analog, but they are being rapidly replaced by digital PBXs, although not necessarily because of the advancing technology. The digital PBX lets customers own their PBX at a low price and doesn't occupy a lot of space.

The structure of the digital voice PBX is similar to the structure of other digital switching systems and is shown in Figure 8.4.

Fig. 8.4. PBX Switching Arrangement

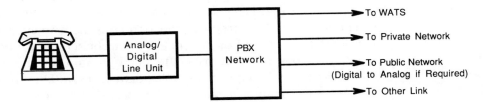

The lines and links are brought to a switching network that operates on a time-division principle to establish a switching path for the desired connection. The network for this path and others is nonblocking: that is, a path always exists or can be obtained through the network for a connection. The line stage, however, may employ concentration, which could cause delays in accessing the network. The number of paths through the network is usually more than sufficient to meet the level of service required by the various installations.

Because the networks for digital PBXs convert the analog signal from the customer into a digital form, it is natural to expect the network to be capable of taking input from data instruments. This input initially is in binary form, and no conversion from analog is required. However, the world is never that simple. The data sets have very high-frequency components that cannot be accurately reproduced with the sampling rate of digital systems. This is where the workstation will appear and provide answers to these interface problems. By combining the voice and data at the workstation in a form suitable for the network, the digital PBX can handle the full spectrum of voice/data.

The architecture for the combined voice/data network will change to accommodate this new service and will be structured as in Figure 8.5.

Fig. 8.5. PBX
Voice/Data
Structure

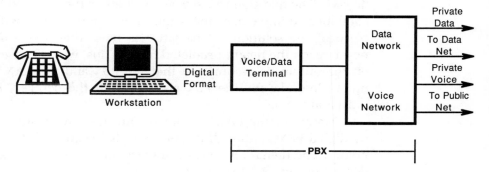

The PBX is responsible for determining whether the user is engaged in a data, voice, or combined connection and for providing the necessary facilities for handling the call requirements. The data call is a packet connection that can use the PBX as a LAN or route to a data network. The voice call is a circuit connection equivalent to current voice connection. Thus a voice bus and data bus architecture are necessary for the PBX. It does not eliminate the need for a LAN in most offices. Few individuals will require the voice/data workstations, but many will still need data terminals and word processors interconnected with other terminals and mainframes.

There are some variations on this arrangement. AT&T, for example, employs a LAN that will connect to a PBX when it is determined that a data network connection is necessary. The connection through the PBX is a circuit connection, although we may see changes to this arrangement in the near future. It also requires separate instruments at the user's location.

Once the PBX is established as the unit for switching data and voice, it also will be capable of handing control interfaces, such as lighting, air conditioning, smoke detection, and other environment controls. The PBX is being established as the intelligence center of the office. LANs also will attempt to seize this market to expand their horizons. In either case voice-to-digital conversion will take place at or near the customer's instrument. Thus the same pair of wires can be used for voice communication and data transmission or both, depending on the requirements of the office.

The LAN as a Switching Device

The ubiquity of the LAN necessitates a serious review of this device as the switch of the future. The recent endorsement of IBM for a token-passing LAN will add to its credibility and may make it as famous as the PC. Most LANs operate with a packet-switching format, and the handling of voice traffic is the challenge. As discussed, voice traffic requires a full-period circuit or a packet method where there are no variable delays. However, the technological evolution of the LAN will permit it, I'm sure, to handle intra-office voice calls with connection to PBXs or other circuit-switching systems for calls beyond the office.

A possible arrangement for a LAN's structure is given in Figure 8.6.

Fig. 8.6. LAN as a Switching Unit

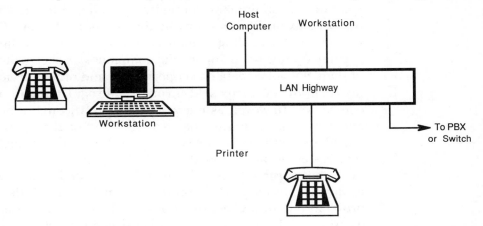

The critical factors for a LAN in this environment would be speed, throughput, and the ability to assign a priority to voice calls. In addition, the PBX or switch may be used for packet-switched calls, leaving the LAN destined for a data network. In this case the connection through the PBX or switch should be a packet-switched call if we are assuming maximum efficiency. Either this creates packet switching with the PBX, or the LAN must have a separate connection to the data network. The latter appears to be the strategy of IBM as it approaches the market with a multitier data network.

There is every reason to believe the LAN will have a prominent position in future telecommunication. Its exact role is not yet known, but it is obvious that it will at least be handling the interconnection of digital devices and will find its way into the home to perform the same function.

Centrex

Although presently centrex can be classed with PBXs, we must comment about its future. When studying the future needs of ISDN together with the current capabilities of a centrex switch, we can easily conclude that here is the switch of the future. The centrex switch offers a combination of the PBX features with the switching capabilities of a central office. For many years centrex fought an uphill battle for market share because each phone was assigned an access charge equivalent to charges assessed to local customers. Hence, until now the success of centrex within North America has been limited. Outside North America it is virtually nonexistent, because few post offices (they control telecommunication in many countries) saw a need for it. The North American centrex market was hurt by the Computer Inquiry II ruling of 1980. Under this FCC ruling, enhanced services could not be offered by the regulated units of Bell operating companies. Packet switching, transport of digitized voice, data base operations, and other services could not be provided by a centrex sold through the regulated end of the operating companies. However, as life would have it, through a number of clever and not-so-clever moves the regulated end of the operating companies ended up with the centrex switches but with restrictions on which services could be provided.

The increased sale of centrex service today is largely from small companies wanting (I believe) to avoid making a PBX decision that could prove wrong in six months. However, buyers may be making the right decision for the wrong reason, because centrex could provide the answer to networking of the future. An increased interest in centrex for the right reason does require a positive response from the current Computer Inquiry III and less regulation on the rates. If both of these studies produce a favorable climate for centrex, the boom is on. Regulation aside, the centrex switch has some attractive features for our ISDN environment. For example, the business that encourages electronic cottages can have its employees as members of the public net while they are members of the business's private net. The latter will require a special set of features and services, and the centrex switch appears ready to provide them.

Centrex allows a customer the freedom sought, and if this switch continues to be regulated the marketplace will seek alternative solutions to these needs. Then the question is "Why did we deregulate?" if the optimal solution to the customer freedom and growth is still regulated and, therefore, noncompetitive.

The centrex switch is ready to compete quite effectively with PBX and central offices. The use of centrex as a viable tool for access to special services, bandwidths, data bases, or other ISDN markets could crush the bypass business because a single unit is able to handle the various networks. Much more will be heard from centrex.

8.11 User Interconnection

The telephone set is a communication link to the world for the customer. The telephone has two functions: (1) signaling to and from the central office; (2) transmitting the customer's information.

For years the telephone has been virtually unchanged. However, during the next few years it will undergo a radical transformation as we start to see "superset," "gigantic set," "does-all-set," and a host of others. This will occur as entrepreneurs enter this arena to "make a million."

A major problem facing the area of the new instrument is the name by which it will be called. I don't think telephone, computer, or workstation will survive over the next few years as new features are introduced on the instrument. Some type of reference to "desk top device" or "desk top communicator" appear more appropriate to the functions it will be performing. It will be interesting to see what the final name will be.

The principal feature of a telephone set will no longer be its reliability but its ability to handle voice and data connections. A secondary feature will be the reintroduction of Picturephone service. As videoconferencing grows within the business world, the novelty of a Picturephone will also grow. People will soon find the service attractive, and the growth will start.

For the packet switch the combining of the voice and data interface is expected to be as shown in Figure 8.7. Note that the customer set contains data terminating equipment and voice unit in addition to a device for separating the two. This device also will, most likely, convert the analog voice into digital transmission. The voice network will have a similar device at the customer site but will use the network to separate and route the data call to the proper network.

Fig. 8.7. Line Interface for Data Network

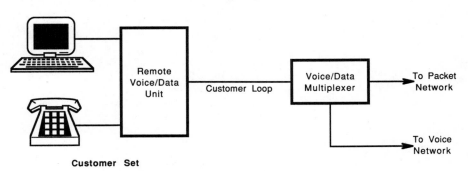

In the voice/data arrangement shown in the figure, the multiplexer determines whether the customer is making a voice or data call. For a voice call the connection is forwarded to the end office or voice switch for processing and the call proceeds just as any other voice call. That is, the customer is provided with a full-period circuit.

A data call is switched through the packet network and the multiplexer caters to the speed of the customer's terminal. This connection would be equivalent to other data connections within the packet network.

The customer interface will be a major battleground for voice versus data networks. Both protagonists are of the opinion that the ubiquitous workstation will determine the network of the future. For example, PBXs are attempting to combine the data structure with the existing voice structure to take advantage of their present position in the business environment. Meanwhile, packet networks are keeping pace by introducing units to compress low-speed terminals, expand the type of terminal interface, and experiment with packeted voice. Hopefully, the future will see a combination network incorporating the best of both switching arrangements.

9

Integrated Services Digital Network

The information age started in 1947 (transistor, information theory, computer structure), and no one noticed; the space age started in 1957 (*Sputnik*), and everyone paid attention. The introduction of the network for information age, the integrated services digital network (ISDN), will probably not be noticed as the average person uses the telephone instrument unaware of the myriad applications available.

The worldwide telecommunication network pushes technology, improves the quality of life, raises the standard of living, and makes the world "smaller." Yet, a servant of telecommunication—the space age shuttle—received more attention then the satellites being placed in space. The satellites are one of the steps on the road to ISDN, but unless something goes wrong with them, they are seldom mentioned. It's as if the post office convinced us that the carrier is more important to us than the mail. Other than some emphasis on education for a few years, the greatest value of the space program so far has been as a carrier for the telecommunication industry. Needless to say, not everyone will agree with these statements, but I'm sure they would agree that there are a lot of "unrecognized" people working to enrich our lives by improving the telecommunication offerings and accesses. The most unrewarded aspect of this effort is the setting of standards to provide worldwide communication.

9.1 Introduction

A general problem with both private and public networks is the need to interconnect the users on demand with disparate and dissimilar devices. Interconnection is complex; it can include dialing patterns, transmission speeds, data codes, languages, and numbering plans. Organizing the network for interconnection has become a full-time effort for groups attempting to standardize protocols and provide reliable flow controls without fanfare.

The advent of the information age has caused the end-user to request a single instrument with one set of instructions having access to many voice/data services. The introduction of new services to telecommunications will accelerate this need and, although experts tell us ISDN is years away, a strong drive is currently underway to set the standards for the products of the future.

One advantage of ISDN from a corporation standpoint is a single administration associated with the network, in contrast to the telecommunication and data processing battles. For everyday users it is a place to bring their requests, questions, and complaints. With the current telephony situation many people don't know whether to call the local telephone company, the long distance company or the local interconnect with their various questions and problems.

The integration of digital transmission and switching is underway based on 64-kb/s channels with potential for both voice and data. The use of these channels for data will permit a substantial increase in data transmission rate compared with present voice (analog) channels. Virtually all data and analog services (with the exception of moving video and very high-speed data transfer) can be accommodated on 64-kb/s channels. As a result the integrated digital network (IDN) will combine the coverage of the geographically extensive telephone network with the data-carrying capacity of the digital data network in an ISDN structure. In this case "integrated" refers to the simultaneous carrying of digitized voice and a variety of data traffic on the same digital transmission links and switched on the same digital exchanges. The key to ISDN is the small marginal cost for offering data services on the digital telephone network, with virtually no cost or performance penalty to voice services already carried on the network.

9.2 Services with ISDN

The ISDN services will range from basic telephone or plain old telephone services (POTS) to a wide range of custom services in video, data, or voice for the public network. Presently, many private data networks are being established using leased communication channels and serving large business organizations. In addition, public dedicated data networks, based on either circuit or packet switching, have been established to provide nonvoice communication for the business community. These networks will merge with or drift toward an ISDN environment.

Multiservice networks have several advantages: (1) It is easier for an administration to manage and control a single network's center rather than several centers; (2) long-term economic justification favors one network; (3) one network can take advantage of fluctuating traffic from different services throughout the day; and (4) higher traffic from one network permits a better bargaining position with carriers. The economic justification may be difficult when the initial cost is compared against the present network cost. However, once the initial step is taken with the network, future planning is easier and major capital requests are not necessary. Similar equipment can be used in many locations and ordered in large quantities. New facilities are introduced more easily into the network, and operational staff are easily trained.

The end-user will have easier access and a simpler dialing plan with a single instrument providing a variety of services. This aspect of the network will take on greater importance as the numbers of subscribers and services grow.

Connections with PBXs require special consideration. Larger installations will provide a self-contained microcosm of new services for the business user, requiring organization-wide interconnection; smaller systems will need immediate access to the advanced facilities of the public network.

ISDN facilities can be introduced at virtually any digital node of the network. At each stage of implementation the ISDN will be the best possible combination of the elements available at that time (Figure 9.1). As shown in Figure 9.1, services can be provided by the local exchange or the call can be transported to the toll switch.

Fig. 9.1. Transport Methods for ISDN

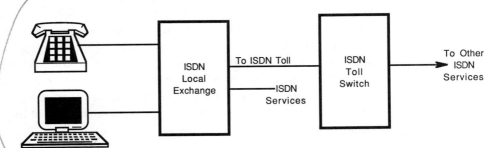

The appeal of the integrated service network is that a number of infrequently used services can be superimposed on the basic telephony traffic, thereby avoiding the establishment of separate networks and the administration of these networks. Everyone has an occasional need for facsimile but not enough to establish a separate network. A brief discussion of some of the nonvoice services presently required for the integrated network is useful.

Facsimile Service

Facsimile service ("facs") is the transmission and reproduction of printed material or graphics at a distant terminal. ISDN will specifically assist facsimile and similar services with standards the various vendors can adopt. Facsimile will be aided by the growth of voice mail and electronic mail. Speed will improve because ISDN provides a clear 64-kb/s channel between locations. See Figure 9.2.

Fig. 9.2.
Facsimile
Options on ISDN

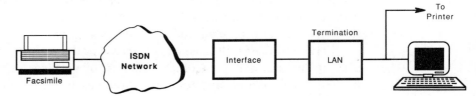

With the introduction of 64 kb/s on the network and the rapid availability of LANs, it is possible to have the distant terminal make the decision on whether the document should be copied.

The originating party calls the party with a distant terminal (in this case a workstation) and transmits the document. The distant terminal can view the document on the screen or can store it for later viewing. If a copy is needed, the distant terminal can access a printer via the LAN and print as many copies as required. You will also notice that there was no special number for you to call, just the number of the receiving party as long as that party or the secretary has a workstation (or word processor with memory) with access to a printer via a LAN or equivalent.

As an interim, CCITT Group 3 has been working on international standards for facsimile that should improve the current situation as vendors adopt these standards. The chaos created by the incompatibility of facs machines has caused many companies and vendors to look to CCITT and other international organizations for assistance with this and other telecommunication services.

Electronic Service

Electronic or telemail service enables subscriber terminals to exchange correspondence on an automatic memory-to-memory basis; communicating terminals are used to prepare, edit, and print correspondence. This service will eventually account for the majority of the correspondence between individuals either at home or work.

Videotex Service

Videotex (known by different names) allows the retrieval of information by dialogue with a data base using standard TV receivers (suitably modified or specialized) as terminal equipment.

Network Impact

Traffic studies for new services show that a digital telephone network can also carry nonvoice services without significant expansion: Even if all business mail becomes electronic and that 90% of telephone subscribers use the videotex service for six minutes per day, the impact on information flow in the network is small, because data and text both have a low information content, in terms of bits per second, compared with voice.

With telephone traffic as the base (100%), the additional traffic could be about 6% for facsimile, 1% for videotex, and 0.05% for telemail. The ability to store and forward text or facsimile at the sending terminal spreads traffic throughout the day and the impact on the busy hour will be minimal.

In addition, the network will be able to evolve easily to incorporate technological changes and enhanced or new services. It will have to incorporate new transmission media such as satellite networks and fiber-optic systems. In the initial stage it will have to interconnect the basic digital network and a wide variety of data networks, such as packet- and circuit-switched data, datel, telex, and other variants of private networks.

9.3 Network Architecture

Digital exchanges are based on 64-kb/s synchronous transmission and circuit switching. The range of application covers local, tandem, and toll exchanges according to network utilization. Concentrators and remote switching units are capable of being associated with the local exchange to ensure end-to-end 64-kb/s transmission. Conversion of the subscriber analog voice signals to digital form is done at the local exchange or by

associated remote equipment. Subscriber signaling at present uses tele-
phonic methods, but enhanced signaling for the digital local area is
under study.

Nonvoice services are digital in nature and cover a wide range using
asynchronous or synchronous transmission. Circuit switching, in which
a connection holds its physical path through the network for the duration of a call, is not suitable for most nonvoice services, particularly those
that involve burst-type transmission of bits. Packet switching does not
allocate a part except when data is actually flowing and is more suited for
transmitting this type of information.

In all networks the user terminal is connected to the network via
data-circuit terminating equipment or its equivalent. The signaling between digital exchanges should be conducted by a separate channel using
a message-based signaling system commonly referred to as common-
channel signaling (CCITT#7 or CCIS).

In spite of the differences between the requirements for voice and
nonvoice services, service integration at the customer level is feasible by
providing special electrical and signaling interfaces for subscriber access.
The use of common-channel signaling and PCM (pulse code modulation)
transmission simplifies the junction and link side of the terminations by
retaining the standard 1.54-Mb/s PCM interface; in addition, a single 64-kb/s
link interface may be required for some data network access applications.

Internal handling and transmission of data characters and envelopes
within the exchange will need to be compatible with internal digital voice
operation. Link transmission economics will probably require submul-
tiplexing of the lower data rates on a 64-kb/s carrier, particularly over
long distances. It is quite possible to provide all the data-type user and
administration facilities within the same integrated switching exchange
or partition them with the special data networks when access is via local
ISDN exchanges.

Telecommunication network architecture is evolving into what can
be called the transport for ISDN, as CCIS and 64 kb/s start to dominate the
switching and transmission sections of the network. The digital format in
the telecommunication network has made these new services available on
a limited basis while the interface standards and the bandwidth problem
for transporting these services end to end are resolved.

For a few protocols the new ISDN architecture will feature the
following:

End-to-end connectivity   Signals will enter the network in a digital
format, be transported across the network in a digital format, and
exit the network in a digital format.

Customer access The customer will be able to access a wide range of network services via a single link and terminal. Present methods provide a multitude of accesses and standards, which makes it difficult to use the network effectively.

Standard digital interfaces Standard digital interfaces will provide for independent evolution of the services on the network and across the full range of terminal equipment. These standards must be defined clearly in the area where the PBX and terminal equipment are emerging separately—that is, they are not being sold as a package. Both sets of equipment must be portable to new environments. In this case the units, PBX or terminals, must be capable of operating with a wide range of products.

Common-channel signaling The network must operate with a standard for common-channel signaling, preferably from the CCITT organization, that would offer worldwide compatibility for the switches and customers. The basic feature, from a customer viewpoint, would be the fast call setup achievable by common-channel signaling. This service is considered vital.

Customer control The customer must be capable of accessing the many networks and network services available, and, once accessed, control the information flow within these networks. This control must be implemented from the customer terminal.

Providing these functions within the framework of the present installed base requires the planning and evolution of the telecommunication network from its present base to an ISDN base. The network will be taken through an IDN stage and then into ISDN. The IDN stage will provide times to test new features, services, maintenance concepts, protocols, and implementation techniques. It also will be the time when political organizations can help shape the ISDN concept.

The basic requirements for an ISDN arrangement are bandwidth, digital transmission, and protocols. Bandwidth and digital transmission have been dealt with, but protocols require further elucidation.

9.4 Protocols

Most telecommunication services in the next few years will be labeled "pseudo-ISDN," despite the fact that these services demonstrate the value of an integrated network. Why the contradiction? One reason is the need for the network to test new features and services. Individuals or organizations must have access to the network without meeting each standard of ISDN. This allows the public to decide if there is a need for these new services. However, during this time the tag "pseudo-ISDN" or "ISDN-like" will be associated with any feature or service not meeting the total spectrum of standards.

Another reason is the slow evolution of the network toward ISDN. Services must be offered during this period, especially in PBXs, and they will be implemented to suit the PBX and not ISDN. Any PBX vendor can provide the necessary interface to implement voice/data service within the system, but technical capability is not the issue. The issue is whether a terminal connected to a particular PBX can be moved to another PBX and still operate. The CCITT view is that a terminal's speed and other characteristics can be compatible on a network level without regard to the equipment interface. From the customer standpoint the end result, universal interconnection and use of the feature, will still be achieved with either approach, and that is the important consideration.

The main impetus for ISDN will be the competitive environment. Being able to demonstrate the ability to provide ISDN in the marketplace will equate to the capability to interconnect disparate and dissimilar devices. Customers will view this capability as a hedge against an uncertain future, thus providing the vendor who can demonstrate these features a tremendous advantage.

The ISDN structure can be broken into four distinct areas: the channel, reference configurations, interface structure, and protocol reference model.

Channel

The channel, which also can be called a link or trunk facility, can range from a two-wire analog unit to a device carrying the highest digital rate between offices. Each channel has been given a separate designation, allowing individual standards to be adapted.

The ISDN channels are

A channel　Conventional analog voice channel, which would use a modem for transmitting data

B channel　Conventional digital channel with 64-kb/s switching for data or voice and will be packet or circuit switched

H0 channel　Circuit-switched channel with 6 × 64 kb/s (384 kb/s) units for high-speed data or image; no associated signaling

<u>*H11 channel*</u> Circuit-switched channel with 24 × 64 kb/s (1.536 Mb/s) units for high-speed data or image; commonly referred to as T1; no associated signaling

<u>*H12 channel*</u> European version of H11 with 30 × 64 kb/s (1.92 Mb/s)

<u>*D-channel*</u> Packet-switched channel with 16 kb/s for lines, 64 kb/s for links; signaling data for B, HO, H11, or H12

<u>*E channel*</u> Packet-switched channel with 64 kb/s for links; similar to D channel.

Reference Configurations

For customer interfacing to an existing switching system, the reference configurations for a user to a PBX, LAN, or network exchange are shown in Figure 9.3. The access is divided into two general categories: (1) dedicated to a PBX or LAN, (2) or connected to a regular switching system. The PBX/LAN interface is where these two networks will battle and eventually merge as they both try to match each other's features. It also is the area where the communication manager and the data processing organization will battle for major stakes in this access. The basic question is: "Do you add data or data access to the phone net (the PBX argument), or do you add voice to the data net (the LAN argument)?"

Fig. 9.3. Reference Configuration for ISDN

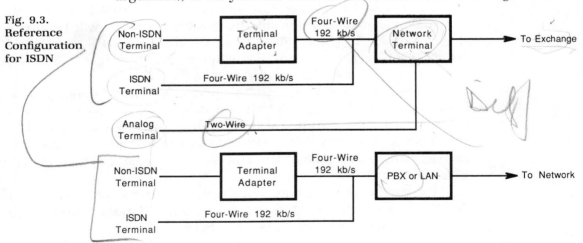

In either case the customer will have the benefit of services and speed. Users of data processing terminals are accustomed to speed, and it must be maintained when voice and data are merged.

The connection to a standard system will have to contend with analog lines as well as digital interfaces. Various options will be available to these customers, depending on the switching system and the level of competition. The technology will introduce schemes to allow an analog line to operate in a similar manner to the digital line. This will permit most, if not all, services to be available to the customer.

Interface Structure

The interface structure will define the method whereby the systems will connect into and through the network for special accesses. This arrangement is shown in Figure 9.4.

Fig. 9.4.
Interface
Structure

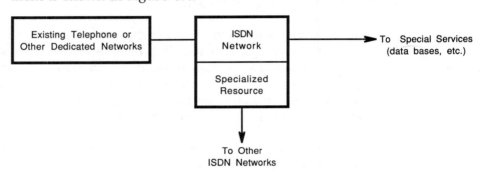

The interface from the ISDN customer to the switch will be 192 kb/s, but the maximum available bandwidth for users will be two B channels plus a D channel (2B + D) or 144 kb/s (2 × 64 + 16). Hybrid arrangements will also be available with the analog channel (A channel) in order to work with the installed base.

The interface world of digital switching systems is presently a 24-channel environment called T1, the North America standard. This will continue to be the standard for the transmission rate for the ISDN, although a great deal of effort will be made to restrict the options. The channel structures can be 23B + D, 24B, 23B + E, 3H0 + D, 4H0, H11, or mixed B/H0.

The maximum bandwidth available is 1.536 Mb/s (the exception is the 3H0 + D, which is 1.216 Mb/s) for these configurations. The interface structure to other networks is being defined.

Protocol Reference Model

The protocol reference model is based on the seven-layer open system interconnection (OSI) model previously shown in Figure 6.3. This model basically is the same one used with the data networks, for which the first three levels are defined by the X.25 protocol. However, there are additional options, depending on the ISDN connection. Information on these options is becoming available through the committees and organizations concerned with ISDN.

9.5 Network Access

Line Access

The first interface for the user is the line equipment to the system. This area originally consisted of a line circuit with supervisory functions. Recent systems have introduced the conversion from analog to digital within the line circuits and the ability to connect to both a circuit network and a packet network. Figure 9.5 illustrates the portion of the system for the line interface.

Fig. 9.5. Line Interface for ISDN

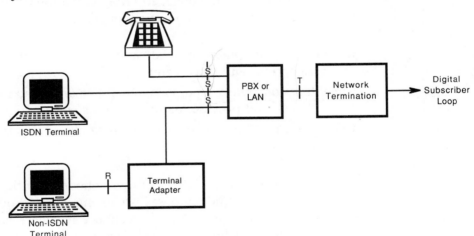

A standard line circuit provides interface to the circuit network and converts the analog signal to a digital signal for transport across the network. There is some question about whether the line also should interface to the packet network. The two main issues in dispute are whether the line performs the conversion to provide the flexibility for packet access, or whether the number of conversions is sufficiently small to warrant the call being switched through the circuit network to the packet network when the service is needed. "S" and "R" are protocol standards for the line interfaces and will provide guidelines for the vendors associated with customer premise equipment and switching systems.

Many volumes will be written on this subject, and we do not intend to dispute the subject here. It is sufficient to say that the data traffic is so small compared to the voice traffic that the switching of the traffic through the circuit network should in most cases be more economical than giving double terminations to each line. For lines where it is known that voice and data are needed, such as featurephone or voice/dataphone lines, two terminations may be justified. I suspect that someone may argue for a circuit network termination only when switching is done through the circuit network, and for a packet network termination when a data connection is required. I don't think this is consistent with the concept of ISDN, but it could be the most practical and economical approach to the problem as we move from a circuit network to an ISDN. When that state is reached, only one connection will be needed.

The line circuit will be capable of delivering 64 kb/s for the voice circuit transmission through the network. The line circuit consists of an analog-to-digital converter, a digital-to-analog converter, and supervision of the customer's line. Any special-purpose line, which will be discussed here, requires these functions, and they are assumed in the description. The featurephone is a type of combination phone and offers

Voice/data   Data phones provide ability to display the data information concurrent with voice connection

Voice/data/compute   A personal computer arranged to interface to the network and provide features such as voice mail

Voice or data   Can handle a voice or data connection but not simultaneously

There are many variations of these phones, but the above represent the major services found on these instruments. A multitude of features are available with these phones if the connecting offices and networks are capable of incorporating voice mail, facsimile, or other modern services.

Conference circuits normally require special connection to the network, depending on the type of conference needed. A videoconference can require a 1.54-Mb/s connection if suppressed video is used. Voice conference requires only a number of ports (or connections) on the network to handle the number of customers to be connected at one time to the conference.

Each line will be connected to the circuit network only or to the circuit network and the packet network. If connected to both, the computer or controls will determine which network the call will be switched through as it goes toward its destination.

Network Access

The circuit network is a time-division network probably working at the T1 rate (1.54 Mb/s) and is nonblocking or nearly nonblocking. Most time networks have low probability of blocking, so for all situations except extreme overloads lost calls can be ignored. The main features that should be in a circuit network are the possible interfaces and the mode of signaling between networks. Some type of common-channel signaling (CCIS, CCITT#7) should be employed because of its speed, security, and ability to access data bases. The interfaces refer to the network ability to connect to other networks, not only the long-distance network of AT&T.

One prime example of an interface need is operator service, which may be more than just special connections or a reservoir for directory numbers. A network should usually have its own operator services, but they should be centrally located to minimize the number of operators. In addition, the operator must be able to access data bases relating to the type of information this service provides.

The packet network must handle X.25 and X.75 interfaces: the first from the line, and the second for connections to other networks. The circuit network and the packet network together with their major interfaces are shown in Figure 9.6.

Fig. 9.6. Voice and Data Interface

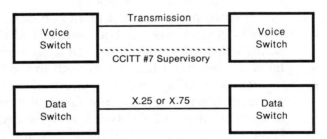

9.6 The Road to ISDN

The reason for the ISDN structure is to allow the telecommunication systems to evolve without worrying about interface to other systems. In other words, ISDN is only concerned with the interface to a customer or a switching system, not with how the system operates.

ISDN can support customer and system services through network configurations. It uses the basic architecture of the ISO (International Standards Organization) with enhancements to support out-of-band common-channel signaling. Within this architectural structure reference configurations are presented along with numbering and addressing samples.

However, before ISDN is reached, the controls and functions of a universal network must be available to permit the daily voice service to continue.

The need for common-channel signaling is evident, and signaling system No. 7 (SS#7) is the response to this demand. The main advantage of SS#7 is the optimization of the signaling in a total digital environment. It is expected that this particular aspect of ISDN will continue to draw more attention as network interfaces to special data bases grow.

The customer interface is perhaps the most important aspect of the ISDN standard. There is no network without the connection between the user and the equipment. This area has recently undergone changes in order to introduce an 8-contact connector replacing a 15-pin connector, and acceptance of this unit is still underway.

The protocol for the data link access is basically the use of the ISO architecture with particular emphasis on the D-channel network interface. The D channel eventually will carry three basic subchannels of information between the user and the network. One subchannel will convey signaling for control of the connection. Another will provide packet-switching service. The third subchannel will be for optional services and telemetry applications.

To accommodate the present network and the evolution to ISDN, an R series of protocols has been established to assist current design. This series will provide the user with a methodology for interfacing to the ISDN service and a base for growth to the ISDN.

Does building a network mean meeting the requirements of ISDN? No. ISDN is a protocol for interfacing the units, and, although this is an essential element of a network architecture, it is possible to develop a good network strategy without ever using ISDN. The reason is that the basic goal of a network is to interface users, services, and features; the protocol is the standard for these interfaces. However, ISDN should be extremely attractive for firms wanting to establish international networks or looking for a good protocol.

Integrated Digital Network

As one of the prime steps towards ISDN, an IDN will be placed in service. The IDN is needed as a backbone network on which services will be tested and imposed. The IDN will switch traffic at 64 kb/s, but more importantly it will serve as an access point and an interface for different services. This strategy will build markets and test technology while permitting the network to evolve toward ISDN. Also, customers with one instrument on the desk will have access to a wide range of services, including electronic mail, data processing, and information processing.

Initially, it is likely that the services will employ separate access facilities, because controls and administration warrant discreet systems. The second phase of this evolution will be the combining of the customer access facilities for services into one network at the customer (or attribute) level. This means that the customer's interface will meet CCITT and other standards for access to the network. It also means there will be networks that will combine these facilities and not meet these various standards.

Circuit and packet switching up to 64 kb/s will be offered for either voice or data with signaling, thus allowing such services as electronic directory or message desk.

These networks will generally operate quite well and interface with other networks for the transmittal of information. The main difference will probably be the ability to transmit at an extremely high rate between a pseudo-ISDN and a true ISDN. Examples of pseudo-ISDN schemes include unique services within a network that the network owner does not want to give up.

Channel Types

CCITT has defined channel types for interfacing with the ISDN, which range from an analog channel to a digital channel with a speed of 1920 kb/s. The important channel from a customer's viewpoint is B, which is a circuit-switched channel for voice or high-speed data operating at 64 kb/s. Data transmission may be circuit switched or packeted on an end-to-end basis (one or the other for the duration of the connection). It is assumed that there is no associated signaling included and that a clear 64-kb/s channel exists.

We may assume that this is easy, because the present T1 units operate with 1.544 Mb/s and consist of 24 channels of 64 kb/s. Unfortunately, there are some checking bits and restrictions on the number of consecutive zeros, which reduce the speed to 56 kb/s. This prevents the present arrangement from providing a clear 64-kb/s channel, the channel most vendors are attempting to provide within their switches to be in a position to say that they are ISDN-compatible. Configurations are being proposed to overcome this restriction, retain the same level of checking, and provide end-to-end 64-kb/s performance.

9.7 Evolution of the Network

Network Access Evolution

The first central office digital switches appeared in 1980. Since then they have changed the environment so rapidly the entire network will probably be digital before 1990. No other switching technology has ever impacted telecommunication so rapidly and so completely. The original digital switches were responsible for the customer's interface and interfaces to other central offices. The current view is that the central office switch will be responsible for extending digital techniques to the customer premise or as close to it as possible. These techniques will permit the customer to participate in the new information age.

Basic telephone service has not changed much in the past 20 years. Most people still view the phone as an instrument for voice communication. They have no understanding of the changes currently going on and absolutely no appreciation of the reasons for deregulation or the breakup of the Bell System. A recent survey indicated most people disapprove of the breakup. Although part of this attitude can be attributed to "don't touch anything that works," I think most of the reason for this unhappiness is that the customer has seen nothing new with the breakup. If anything, the only noticeable change is a higher, more complicated phone bill.

Before discussing network planning, we will examine how the network is expected to change in the next few years. The network must evolve from its present state to the ISDN without creating any interruption in service while introducing new technologies and customer services. Initially, there may be a multiplicity of terminals performing different functions (i.e., voice, data, video); once this is achieved, multifunction terminals will appear. This allows the customer to decide the time frame for this evolution. The first phase will look something like the arrangement in Figure 9.7A. Notice that the voice circuit is switched, whereas the data circuits are hardwired through the switch; that is, the circuit has only one termination it can see.

The growth of this service will influence the second stage of this evolution, because a heavy demand for it will necessitate switching data connections to cater to combined voice/data terminals and new data networks, and the third stage will preempt the second stage. If the demand is light or normal, the second stage of evolution will have the network switching data and voice connections from different terminals. This is shown in Figure 9.7B.

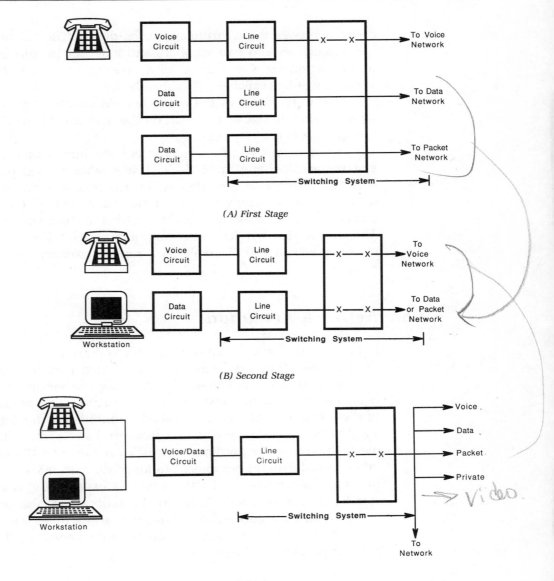

Fig. 9.7. Stages
of Network
Evolution

(A) First Stage

(B) Second Stage

(C) Third Stage

This arrangement will allow calls to be routed to any network to which the switch has access for data or voice. This step would allow workstations to have access to a data network or a voice network, but would require two terminal connections or two lines. The teleterminals available from Bell, Northern, or GTE use this approach, but they operate in an analog format to conform to the current analog network. By having this type of terminal connected to digital networks, the customer, in most cases, perceives the environment to be digital. The customer may have two directories and two numbers and may not be able to do as many multifunction tasks as with a fully integrated network. However, most services will be available with this scheme.

The third phase of the evolution will combine the services at the workstation into one termination on the network. Figure 9.7C illustrates this arrangement.

The network is now a fully integrated information transportation system for voice, data, facsimile, and video, although video could be slow-scan. The service is now defined by the terminal rather than by the network: that is, plain voice service (telephony), or plain data service, or a combination of these services is available at the customer's discretion. The workstation will have only one number, and, if the network is private, the switch will probably be a centrex exchange or a variation of centrex.

Data Network Evolution

Data networks have grown considerably and are now an integral part of the customer functions. Any method for incorporating these networks into ISDN must do so without interrupting the service. The growth of data networks is in interconnecting them to move data from one network to the other. The ISO has established standards for this interconnection via the OSI as a reference model. IBM also has an interconnection standard, known as the systems network architecture (SNA), that allows other manufacturer's equipment to interface with IBM's. As the capabilities and configurations of data networks expand, the need to provide standards and interfaces between these networks also increases. The configuration in Figure 6.2 delineates the various interfaces between data networks and the protocols, as well as other networks, that they must meet to provide the interface.

The Compression Years

Perhaps one of the most rapidly evolving data transmission techniques is time-compression multiplexing for achieving 56 kb/s through the existing wire pair. The Bell System originally offered a Dataphone digital service (DDS), which was an end-to-end data service but required a special line from the customer to the serving central office. A better answer is the attempt to use the pair of copper wires now connecting the customer with the office and to transmit 56 kb/s over it. The technique employed is to have a transmission rate to the customer substantially higher han 56 kb/s (for example, 144 kb/s or 160 kb/s) and to compress the customer-to-office data on one part of the higher rate and the office-to-customer data on a separate part of the higher rate. The Bell System recently announced a time-compression multiplexing technique for this service called circuit-switched digital capability (CSDC).

Northern Telecom Inc. also announced a similar service, called circuit-switched digital services (CSDS), for their switching systems. Both provide 56-kb/s data service, but Northern's is a data-only service, whereas Bell's is a voice and data service. Northern uses the time-compression technique, called *ping-pong*, for deriving the 56-kb/s data. Both systems rely on a nonloaded loop to incorporate their service, although there may be some difference in distance. This technique has several names, including ping-pong or burst switching, but is really nothing more than a compression or multiplexing technique. The interface to the customer is illustrated in Figure 9.8.

Fig. 9.8. Switching Arrangement for CSDC

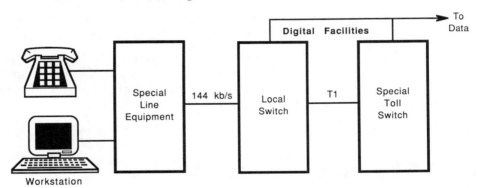

The switching systems at the office must be capable of transmitting at the 56-kb/s rate and routing the call on digital facilities to the destination or to a special routing system for completing the call. These facilities can be fiber optics, digital radio, or T1 lines. There also are options on whether data and voice or data only are available. The argument for the latter is that if the terminal is to be used for data, why mess it up with voice? In either case this technique is the telecommunication industry's answer to the need for high-speed data to special customers. When the network is converting to ISDN, it may satisfy that requirement.

A variation of this scheme is presented by data-switching vendors and called local area data transport (LADT). It is normally used with slow-speed terminals and is based on a technique called data over voice, where the data is imposed on the connection during the idle periods of the voice. Again, it can use the local loop, and the analog portion of the connection is passed to a voice central office while the data portion is combined with other data to form a 56-kb/s circuit.

The multiplexers used with LADT and other schemes are becoming an integral part of the data and voice networks and therefore are likely to expand their features and functions during this period. For example, they could be arranged to provide multiple routing in case of a failure in a higher-speed multiplexer or in an office. Also, many multiplexers can expand to handle more interfaces and serve as converters between optical and electrical transmission.

Some types of digital multiplexers are listed here, although options are available.

Fixed multiplex structure   There is no concentration, and the channels are preassigned.

Semiflexible structure   There is no concentration, but the channels can be reassigned when changes are required.

Flexible structure   There is concentration with the ability to assign channels as required; no standards for this unit.

Intelligent multiplexers   They appear to be the units being designed today, which allow the user more flexibility and interfaces. For example, they can route to different host computers in relationship to the user's service.

9.8 Advantages of ISDN

ISDN will offer several advantages to users and designers of communication networks. For a few protocols a wide variety of services will be available. Multiple services will, via integrated access, allow the sharing of a digital conduit and thereby minimize the cost impact of these services.

The so-called office of the future is something everyone talks about, but no one understands, and it is really a plea for simplifying the many services that are here or on the horizon. My definition of the office of the future goes back to a statement made by Theodore Vail years ago: "One policy, one system, universal service." For the office of the future: "One instrument, one number, universal service."

ISDN also offers a separate channel, the D channel, for transporting information between the customer's equipment and the network. This control permits the ISDN to treat the bits of information from the customer's equipment uniquely. But ISDN depends on having an all-digital network, so it will take years before the network is converted.

An important feature of the ISDN and the separate channel for transporting information is its ability to transmit the calling customer number to the destination, which previously has not been possible. This ability will allow a multitude of new network features, including selective call rejection, distinctive ringing/call waiting, selective call forwarding, customer-originated trace, automatic callback, and many others. Customers will find many uses for these new features, although they are basically unaware of them now.

Centralized data bases will allow a storage point for customer directory information and services to which the customer is entitled. This arrangement separates the switching function from the administration of the calling and called numbers. Although present stored program controlled (SPC) systems are able to separate the customer's number from the physical location on the switching system, they are not equipped to administer the number separate from the switching function. This separation will be the next major and important function switching systems will be required to perform. Through CCIS or CCITT#6 the telecommunication industry was on its way to introducing this function. A temporary roadblock occurred when deregulation was introduced, and the need for CCIS was questioned. Also, if there is a need, who would be responsible for the service? The service will require the cooperation of AT&T and the BOCs, since AT&T would be the transport and the BOCs would be the local switch interface. An additional problem is that any arrangement between these organizations may be viewed as a violation of the recent decree.

However, the need for the function is evident and far outweighs the politics of the situation. Therefore the feature will be implemented, and interfaces will be defined to allow the distinction between voice, data, or special-feature calls.

The workstation or customer subset will consist of the customer equipment and a network termination. The network termination will provide the proper interface to the network and address the network with information on what services are being requested. Hence, the network termination is aware of the services possible from this customer and is programmed accordingly.

10

Network Management

In the past two chapters we have considered some of the current problems connected with new networks. In this chapter we will still consider other problems that arise when we reflect on the most pervasive feature of the network we want to build. Here, more than elsewhere, the cleavage between the transport of a wide-bandwidth call from coast to coast and the pragmatic problems of a network will be apparent.

10.1 Objective

A communication manager, whether managing two private switches or all the switches for one of the *Fortune 500* companies, must have a plan for centralizing the support functions. Over the life of the system, more money can be saved here than in any other area. Support functions should be responsible for the installation, maintenance, and administration of the entire network. It is not necessary that the people working on support functions actually install or maintain each switch. Normally, many of these functions are best subcontracted. But the responsibility for the overall operation resides with the center. This allows a single focus point for the network functions. It does not preclude many of the functions, especially in large networks, from being performed at the sites. However, the hierarchical reporting structure and the information flow must focus on the center. The individuals should report to the composite center to share information.

An objective of any network plan is to provide overall quality service at reasonable cost or investment. With the expansion into digital and the increasing sophistication available from networks, centralized maintenance has become not only desirable but also economically feasible. Several management tools are necessary to accomplish centralized management (really a better term than centralized maintenance). In addition, a clear understanding of the objectives for the center is needed. To achieve uniform service throughout the network, we want the overall plan to evolve toward software control of the maintenance and administrative functions within the switches and the network.

The network management function is the real-time surveillance and control of the call-carrying capacity within the network. Network management is usually thought of as controls that are introduced when the network is in an overload condition or when there is a major fault in the network but there is an ongoing requirement for this type of management during daily operation. Network management, for example, could take advantage of the fact that traffic volume between locations is a function of the time zone for these locations and use some of the links for different traffic or functions when the load is down. Also, during periods of catastrophic failures, network management is responsible for deciding which traffic should be routed through the network. A surveillance responsibility is inherent in this area to prevent unauthorized use of the facilities and to ensure the proper grade of service to authorized users.

10.2 Key Functions

Associated with network management are four key functions that must be understood in order to meet the objectives for quality service: control, maintenance, management, and administration.

Control refers to the continuous operation of the switches and the devices that interconnect the switches. The devices interconnecting the switches are normally referred to as transmission facilities, but with the advent of satellites and fiber optics, *pipeline* is a more descriptive term. Also, fiber-optic facilities are capable of providing controls that can be centralized quite easily. It has been said that telecommunication will be reduced to electronics and photons in the information age; if so, then the controls must be electronics and photons. This includes items such as the routing approach for voice or data, digit analysis, most economic route, and any other item associated with the transport of the call across the network.

Maintenance includes the activities related to the routines necessary to keep the network in good shape. This includes programs to place test calls throughout the network, run diagnostics, correct problems, and perform scheduled audits. It also includes the placing of calls to simulate problems mentioned on trouble tickets. Certain routine maintenance procedures must be performed at the switch, but more and more of these are being centralized, where manpower can be saved and a network approach to problems can be achieved.

Management (or monitoring) is not part of the normal call flow but something implemented when an unusual occurrence, such as route failure or an unexpected interruption of a call or session, takes place. Many management devices can be incorporated into the routes or switches in anticipation of certain failures. Other devices are centralized to provide for the monitoring of the network activities. Certain problems can be anticipated through the use of a network status board, which indicates the status of the various routes within the network. In addition, management refers to billing and traffic administration routines needed with the network. Certain discrete data elements are required from each switch to derive billing information and to provide analysis of the overall routing needs of the customers.

Administration is the daily functions performed to keep the network and the switches current. This includes the initialization of new users, the termination of users, number changes, and other routines necessary to keep the operational status of the network up to date. This function presently is associated with individual switches, but will become centralized as more data links become available. Administration can also do the updating of the directory. This function presently resides in many strange places within a corporation. Administration has the information necessary to make updates and has access to data bases for collecting, collating, and printing the information. This function requires terminals that can access switches for the information, a large data base for updating, and a method to present the directory when requested.

10.3 Network Control Center

Computer hardware and software packages can provide a network communication center for various services. A network communication center is really cost-justified because it can restore a node without a maintenance man at the site. The center must have this capability, or the rest of the services are useless.

For networks with a few switches the typical approach is to use the functions within each switch. The billing and traffic administration functions are polled or passed to a data processing center for collating and summaries. As the network becomes more complex with a multitude of switches, the need to centralize functions becomes more evident. The obvious advantage is the manpower saved during off-hours in the maintenance and administrative areas (the exception is where a data processing center exists at each switch). A better understanding of the network routing is possible with centralized controls, and this can lead to greater economy. An example of a network structure with a centralized network control center is shown in Figure 10.1.

**Fig. 10.1.
Network
Management
Center**

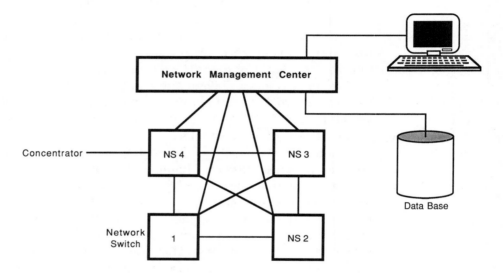

The control center has a connection to each of the switching systems. In addition, the status of the facilities between the switches can be monitored if those facilities are equipped with data links and proper monitoring units. The control center provides around-the-clock monitoring of the switches and a level of security needed in the current environment.

Center Services

The center can handle different functions, and the choice depends on the network requirements. Some functions are operational and maintenance functions, line and link tests, transmission tests, traffic (including service observation) administration, network management, billing functions, and technical assistance.

With SPC systems, it is possible to obtain real-time analysis of the system together with trouble detection. This information can be transmitted to the control center through the use of automated procedures and special lines. The major advantage of this approach is to maintain the network's equipment with the individuals who are most knowledgeable of the various switching systems and transmission facilities.

The line and link testing ensures that these devices are accessible to traffic and are meeting certain transmission criteria. Link maintenance procedures normally involve manual operation conducted from a test position as well as fully automated procedures. Tests of digital links should include the checking of slips, bipolar problems, and out-of-frame conditions in addition to tests required by the administration. Most line testing is performed on a mechanized basis, but arrangements can be implemented to centralize the results.

Transmission testing should be performed periodically, and the results should be logged for comparison with other tests. This ensures uniform transmission throughout the network. The listener's echo is one of the major transmission criteria in a digital network where the customer loop is two-wire and the interoffice facility is four-wire.

Stored program controlled systems can store information on traffic quickly and accurately. This information should be passed immediately to the control center for analyzing and, if posible, the network configuration should be dynamically adjusted during periods of unfavorable traffic conditions. The traffic information in these cases is broken down into two major categories: information needed for dynamic changes, and information needed for determining long-term trends and load situations. Because modern systems can present so much information, the information needed for overload or traffic trends must be judiciously reviewed to avoid an abundance of redundant data.

Center Controls

Network management is the recognition that the individual switches are not autonomous entities and that customer service is determined by what is occurring on the total network. Network management depends on the traffic information previously mentioned, but interpretation will be different. The main objective of network management is to complete as many calls as possible under adverse conditions. The term "adverse conditions" is used because we assume that the network can complete as many calls as possible under normal conditions without network management controls. Some conditions that necessitate network management controls are as follows:

- A switch has an unusually high congestion, and it is necessary to prevent this congestion from spreading to the network.
- High network congestion exists, and calls to the network are reduced by eliminating alternate routing, thereby only permitting first-routed traffic.
- Introduction of reroute procedures where congestion can be reduced by using time-zone differences.

Call routing through the network during catastrophic failures must be predetermined and dynamic for timing considerations because the switches may be unattended during these failures. Also, the controls may be able to anticipate the overload and immediately change the routing patterns. Network management controls during these periods are

- Special register and sender timings
- Cancellation of alternate routing
- Accelerated timeouts
- Priority to terminating traffic

The important aspect of network management is to determine exactly what controls are available within the switches and to formulate an overload plan based on this information. Also, an initial test of these controls during a convenient time is recommended.

An additional function to consider at the center is the collection, processing, and distribution of billing data. Although not necessarily associated with a control center, this function can best be handled at the center. This data can be transmitted to the center in the same manner as other data, and thus, the center serves as a convenient place for this information. The center collects the data for passing to the data processing equipment. Other customer-related functions, such as recent change, may best be served at the center. Recent change is the updating of a switch whenever a customer is added, deleted, or incurs a class-of-service change. If the directory is handled by the center, it should also handle recent change.

The function of technical assistance at the center is to perform trouble analysis and assist the less experienced maintenance people at the individual switches. Obviously, the most experienced personnel are operating the technical assistance center and, in addition to troubleshooting, can frequently make software changes to the switches, which the vendor may not be able to do for economic reasons.

10.4 Major Network Services

The implementation of a centralized facility is a difficult and time-consuming task. Because the tasks can be performed at the local switches, it is often necessary to have objectives for the center and to prove that these objectives, implemented at a center, can provide economic benefits over local implementation.

Some objectives for the center are

1. Formulate a standard network operating plan.
2. Define the information flow among the various systems and plan the automation of the communication between the switches and the center.
3. Clearly define the services to be performed at the center.
4. Develop service criteria for the services at the center.
5. Develop protocol and input/output standards for interfacing between switches and between switches on the center.

A detailed description of the services normally provided at a network control center will be given to ensure a common understanding of the service and its operation.

Service Order/Recent Change

The need to update the switching systems as quickly as possible has led to a centralized service order that can take the information to the system in what is commonly known as a recent change format. Recent change is the information the switching machine needs to update the customer data.

Links would have to be established to the various switches if this function is to be centralized, which is the preferred way to operate. Combining service order and recent change will not only improve operations (less personnel, quicker response) but will eventually allow for an electronic directory. An electronic directory is possible because all the information can be associated with the order and stored in a separate data base, and this data base can be informed when the information is incorporated into a switch. This places the directory in a common data base, and it can be used for printing the directory or providing access to it via the network. We want our attendants and operators to have access to this information and, if successful, to extend it to the user of the network. I would guess that this will be a common network feature in the near future.

Message Detail Recording

Message detail recording (MDR) is billing data collected about a call on a real-time basis. Whether the bill is actually used to collect money is not as important as the need to have information on usage of the network and destination of the calls by various organizations. The information is normally capable of being placed on magnetic tape or outputted to a centralized recording center. Reports can be displayed on a CRT or printer.

The message recording requires the calling customer, the called number or destination, the call start time, the call completion time, the call duration, and information on the links used in the call. Some of this information may vary slightly from system to system. From this information the network center can provide various reports on customers or type of call, but the major reports normally are generated by the data processing center where the information is processed.

Billing data supplied is as follows:

- Called number: normally 7-digit or 10-digit
- Calling customer number: 7-digit
- Date and time of day
- Duration of conversation
- Optional: bandwidth used

There are variations possible on this data, but the manufacturer should be able to meet these requirements.

The traffic administration function is included here because, though not related to detail recording, the information is gathered on the same tape and both end up at the data processing center for sorting and collating. The traffic administration information may also be used by network management to make decisions on reroutes or overloads. There are three categories of traffic administration functions: traffic data, network capacity and administration, and link requirements.

Typically, traffic data can prevent inappropriate use of the phone system and encourage economic use of facilities. The first is an obvious requirement. The latter is the result of deregulation and the introduction of numerous carriers with different rate structures. There are a number of rate structures already, and it is an impossible task to attempt to keep track of all carriers at all times. However, for a particular application, certain structures can be analyzed and advice given to the user for assistance. This analysis should be performed on the places most frequently called by the people using the network. Billing data is necessary for this analysis. A billing system can greatly assist in providing the information necessary to distribute the cost to the various departments properly. This method of allocation can be used to validate the monthly toll billings from the carriers. The benefits of traffic data include employee management, expense control, and cost allocation.

Network capacity and administration consist of information necessary to take corrected action during unusual situations and to detect expected or unexpected shortages so that additional equipment can be purchased before service is impaired.

Link requirements are used to determine routing patterns in the network and where new links are required to meet the changing demands of the customers. Eventually, if the network is large enough, a model is useful for optimizing the routing patterns and link sizes.

Traffic Data Analysis

The traffic data analysis package collects call count and usage data for items within the system that are traffic sensitive. The information is necessary for the instantaneous indication of overload or trouble, to determine whether the present equipment is meeting the service criteria established for the system, and for long-term trend analysis. Many would consider this the most important package at the network center, but many systems are becoming less traffic sensitive, and eventually most information related to the network will be obtained from the billing package. That does not diminish the importance of the traffic data, especially when the network is going in service.

Failure Detection and Alarm System

The network communication center should be capable of monitoring each switch within the network together with the transmission facilities between the switches and, if a major problem arises, to be alerted via an alarm from these units. Alarms typically are classed as major and minor. Minor alarms are those that are turned off and left for routine maintenance, whereas major alarms require immediate action. In addition, alarms can be made available to monitor air conditioners, power, doors, and other actions. A log of these alarms must be maintained in order to trace recurrent problems.

Trouble Log

Three major sources of trouble reports should generate a trouble log at the center: reports from centralized testing, alarms or other reports from the various switches, and customer complaints. The reports from centralized testing are normally accumulated over the testing interval and printed for action that day. The alarm reports require immediate action and should include what action was taken on the summary. The customer complaint also requires immediate attention, and an indication of what service was performed or corrective action taken should be on the form.

The system should summarize the result from trouble tickets into a suitable report required by the user.

Centralized Testing

The testing of all links within the network is a vital feature of any system, and the network must initiate this testing from a centralized point. This is not to say that the center is doing the testing, but that it is capable of interfacing with a link testing unit and storing the results for action.

Automated Directory

The automated telephone directory provides an on-line source of information for any user of the network. The present compiling and printing of directory information is a long and laborious process, with the information being outdated before printing. The automated directory avoids this by providing an up-to-date listing, by alphabetical listing or by function, of employees within the company.

The basic directory information should consist of the employee name, title, department, location, and both the public and private network numbers associated with the individual. The system should also provide a formatted printout of the directory for formal publishing. This is a valuable feature for a normally chaotic function.

Other Services

Other services that are highly desirable and that would naturally fall under the responsibility of the center, but are not absolutely necessary to its operation, include

- Electronic message center for distribution of announcements throughout the company
- Automated inventory system for an on-line, accurate, and current record of equipment
- Call trace to provide information of each link used in the connection for routing analysis and trouble detection
- Load control within each switch
- Link monitoring to determine when rerouting or no alternate routing must be implemented (is tied in with load control)

Other services are offered, and new ones are being added daily. Some of the services, if on-line, could use a large amount of memory. Inventory would be an example of this type of service.

The interface from the switches to the control center is a sizable investment and a complex arrangement. The billing data can be forwarded manually or polled via a centralized device, which in turn would interface to the data processing center for collating and summarizing the information. The preferred method is polling, where there is sufficient time in the switch to pass the information. For the switch alarm a dedicated facility is highly recommended to provide immediate response.

10.5 Testing

Line Testing

The center also serves as the place where the routine testing of lines, links, or common control is initiated. The advantage of having these functions located at the center is to allow trouble reports to be acted on at the center and to permit multifunction individuals to be available to handle the whole network during prime and off-hours.

Effective line test equipment is normally inserted, automatically or manually, between the customer line and the switching equipment. Its function is to test the customer equipment in the direction of the customer line or workstation and to test the switching equipment as if it were a demanding customer in the direction of the switching equipment. This latter function includes placing calls through the network to special lines to determine transmission levels, analog or digital, in the network. Service criteria for this test equipment should be more demanding than the customer, so problems can be detected early.

Line testing also demands that the faults be localized because many kilometers exist between the test equipment and the actual line, and faults can occur at any point in the line. The testing equipment should also be capable of determining whether certain line attachments can meet the protocol standards and service criteria of the network. These criteria are helpful in deciding which instruments to connect to the network or switching system.

The line testing should access the customer's loop when a trouble report is received and verify the conditions. This verification requires numerous tests and computer analysis of the results. The information is made available to the person running the test for action. The equipment normally is designed to be compatible with certain switching systems; modifications may be required to achieve a common interface between several systems.

Common interface or common language, as it is sometimes called, is the ability of the maintenance personnel, administrative unit, or individual operating on different systems to input one set of commands and have the supporting computer translate them into input that each system can handle. This is a key feature in any service center.

Link testing is more complicated, because there are analog and digital links, and under them are a variety of options (synchronous, asynchronous, loop, E & M, two-wire, four-wire), and all require routine testing to detect trouble before it is noticeable to the customer. We will explain link testing briefly, using only two units. We will use the term "trunk" instead of "link" for the facility between offices because this term is imbedded in many definitions.

CAROT (centralized automatic reporting on trunks)   The system that performs an end-to-end routine or demand test on trunks within a certain area.

ROTL (remote office test line)   The device CAROT seizes in the office where the tests are to be performed.

The basic arrangements for CAROT are shown in Figure 10.2. CAROT will dial the ROTL equipment located at the office under test, and

ROTL will connect the necessary test equipment to the connection. The trunk, whether analog or digital, will then be tested, and the results will be passed to the center. In an all-digital environment we are looking for slips or out-of-frame conditions; in an analog setup we look for noise, transmission levels, and such things. The ROTL together with a responder can also test one-way outgoing links.

Fig. 10.2. CAROT System

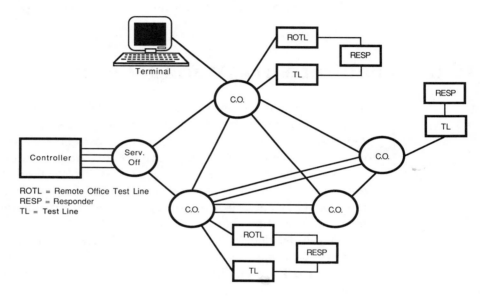

CAROT also will analyze the test results and compare them with predetermined limits in the data base. Tests that exceed these limits will be reported, and statistics on the results and frequency of tests are also kept.

Switching Systems Diagnostics

The modern switching system has the ability to conduct routine automatic operational tests and certain link testing on its own. The typical results are the removal of certain links from service. The system should forward this information to the center for collection and action, if required. The information can vary from a trouble report on one link to the condition that the entire system has gone out of service. The center must be able to respond to these messages and decide whether immediate action must be taken or whether the condition can be corrected during routine maintenance. The center and the switching system must also be equipped with a software dictionary of the likely faults. The dictionary translates the system information relating to a trouble condition into a card or cards location so that the faulty equipment can be replaced. This information is also very valuable in collecting historical information on faulty cards or components.

The amount of information collected at the center must be controlled, or else the center will be inundated with paper.

10.6 Local Network Management

The emphasis is on centralized management for network functions, but that doesn't mean local controls are not necessary or, in some cases, more economical. A few words about local controls are necessary, because many vendors feel they have an advantage in this area.

Any telecommunication magazine is filled with ads for the ultimate equipment to provide proper management for the network. However, most of these ads are promising equipment for a unique service within telecommunication, not for the entire network.

We must decide what equipment is going to be managed. Assume that the total spectrum of telecommunication will be covered—that is, data networks, including front-end processors (FEPs), voice networks, including the PBXs, and any interfaces between the switches and networks where they are used for video, facsimile, or whatever. If network management covers only certain functions, the vendors might have a package for the requirements.

What we don't want is a variety of local systems put together by bits and pieces with multiple data bases, duplicate alarms, and unique command streams. In other words, we have to design a blueprint for the local management system just as we design the network—by taking an architectural view of it before any implementation occurs.

Local network management is shown in Figure 10.3. Each center has its own controls independent of other centers. It is, however, a good idea to have essentially the same equipment in each center, thus allowing us to move people from one center to the other without additional training. The local network management is advantageous when locations are limited within the network, when each location is large, and when each location has a large mainframe computer (or equivalent) requiring around-the-clock coverage. In other words, we are going to train the computer operators to maintain and operate, if necessary, the telecommunication system at least during off-hours, if not full time. This has the obvious advantage of placing the telecommunication requirements under the information management department.

Fig. 10.3. Local Network Management

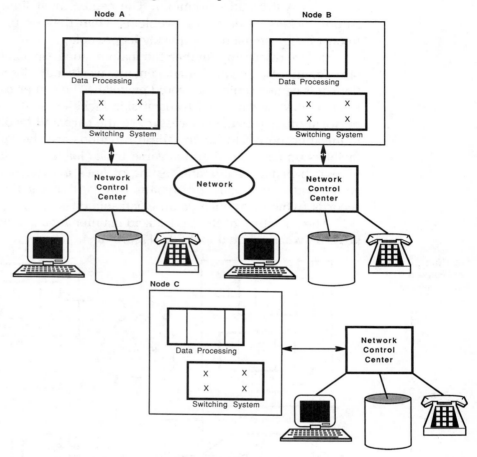

The equipment purchased for the network management in this case is attainable from the switch manufacturer or from an independent vendor with equipment compatible with the switching systems. The determination of whether a local or centralized management system is proper is strongly influenced by the data processing operation within the company.

10.7 Centralized Management

The purpose of centralized maintenance or management is to provide a high quality of service with a minimum investment in manpower and training. With the expansion into digital and the increasing sophistication available with networks, centralized maintenance is not only feasible but highly desirable, because manpower for operating and maintaining these networks will be at a premium. The center must limit the number of trouble tickets by detecting problems when they occur and responding to customer complaints as quickly as possible.

Before discussing further the maintenance aspects (the most important function) of central management, we briefly describe other functions this center could and would perform. The center can interface with the network switches and transmission facilities, so it is ideally placed to perform other functions necessary to the care and feeding of a telecommunication network. Among these functions are the optimal use of the facilities under normal and adverse traffic loads, entering and updating of switch orders, automatic testing of lines and facilities, ability to reroute traffic manually or dynamically, and recording and updating records (which may include an electronic directory).

The structure of the centralized maintenance and control center for large networks should look like Figure 10.4.

Fig. 10.4.
Centralized
Management

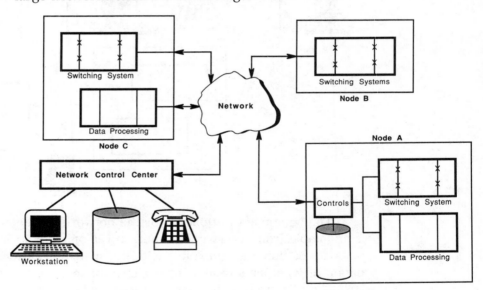

The structure shown here would not apply, for example, to a small office serving one or two private switches. For that application a PC could implement the functions needed or the switches may have the ability to remote the maintenance and I/O (input/output) ports. For slightly larger configurations the interface could be directly to the network console without the distribution processor. The advantage of the distribution processor is its ability to interface to any switch at any speed: in other words, it serves as a protocol interface and a speed converter. Placing the protocol interface in the distribution processor means that the test units and network control do not need to be changed whenever a new switch with a different protocol is added to the network. It also allows the links to the various units to be high speed, so it serves as a multiplexer from the switches.

The emerging network offers new opportunities to centralized maintenance, including better utilization of resources.

The operation and maintenance concept must include

- Maintenance procedures, including efficient aids for diagnosing and monitoring
- Administrative control and effective billing
- Resource utilization, including remote control to minimize the personnel located at the switch

The operation and administration controls of a system must be recognized early so that incorporation within the system is readily achieved. This is especially true in today's environment, because the demands to incorporate many new items are constantly arising, and they must be introduced with few disruptions.

System controls should extend to the handset or workstation, where, from a user's viewpoint, the network begins. If this area is not working, then the network is not working. Telephone operating companies are only concerned with testing to the customer's instrument, because the customer owns the instrument. This is not so in a private network where operation of that line is paramount to the operation of the system. The line is further complicated by the analog and digital environment that will be seen during the next few years. Both must be tested, and both require different test procedures.

The resource strategy for the network should include route control as well as most economical routing.

Other features being touted for administrative control that have caught the fancy of network administrators include automatic directory system, common data bases, and equipment inventory.

The level of service to the customer is one of the most important aspects of administrative control; the types of customers together with the different services required in the telephone network complicate this criterion. For example, a voice customer may tolerate a delay in switching a call destined for a foreign country, but a high-speed banking translation will not tolerate the same delay. We would like to switch both services over the same network; therefore the same service criteria for both calls are necessary. It is expected that the more stringent of the services will prevail, in this case the banking translation service criteria.

Centralized operation of the network and the switching systems provides a common data base for the administration's organization. Billing systems, sales order systems, network maintenance systems, and inventory control systems can all be located in the network control area. The advantage of this centralized control is the effective use of personnel. Billing systems and directory systems are also naturals for this area. The administration of directories for voice networks, data networks, and telex networks is becoming an increasingly troublesome duty. The computerization and centralization of these functions will reduce cost and improve service.

10.8 Control Center Interface

The control center interface, including billing, maintenance, and traffic data, constitutes a sizable investment, and care must be exercised to ensure proper operation of these units.

The billing data can be forwarded or polled via a centralized device, which would then interface to the center computer for collating, summarizing, and billing. The preferred method is polling, because a link between the control center and the switch is not occupied full time and, assuming sufficient storage at the switch, can be performed at night when other traffic is down. At the center a standard interface would be a collector of data for compressing, if required, and passing to data equipment for processing. An option does exist for the computer at the center to process the data, depending on the number of switches and the load on the center.

A sketch of the two configurations is shown in Figure 10.5.

**Fig. 10.5.
Control Center
Interface**

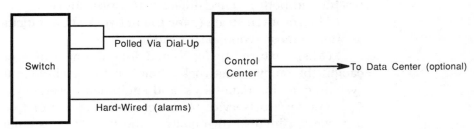

From a maintenance standpoint the service typically can be polled to gather the required information. However, the switch alarms are brought in the center on dedicated facilities because a major alarm must be responded to immediately.

11

Planning a Private Network

We are now ready to discuss guidelines for networks. The problem is to perceive how a network exists for its environment and how the pieces fit together. We will start with management's insatiable appetite for information throughout an organization and evolve toward general characteristics and features of the network.

11.1 Need for a Private Network

Business is becoming more dependent on the flow of information for its daily operation. The key to this dependency is a reliable network interconnecting the units of the business. As this dependency for information continues to grow, the networks and other units associated with providing this information will also grow.

A new concept of the corporation is developing, which will have a tremendous impact on how business is performed. No longer is the operation of the corporation focused on a large downtown building where the various functions are located. The modern corporation is becoming a widely dispersed group of functional units tied together by a telecommunication network and often a computer network. In this environment the operation of running the corporation is focused on the computer and communication networks' ability to store and make information available at a moments' notice. Network communication is essential to the interorganizational information systems and the operation of the organization.

At the local level most manufacturing and assembly operations will function from a network that will interconnect the factory and the office.

To this end, General Motors has undertaken a major development to control their suppliers and integrate their offices and factories with a unique protocol known as MAP (manufacturing automation protocol). This effort is reminiscent of the structure Henry Ford built by the River Rouge where the suppliers were located along the river and any part or raw material could be transported up the river on extremely short notice. This innovation set the standards for manufacturing in the twentieth century throughout the world. GM, in realizing that their production and their suppliers production are controlled by various mainframes, is attempting to create a River Rouge network with MAP, that is, tying the manufacturing and their support groups to one data network. The jury is still out on whether MAP is the answer to the 1980s production problems, but it does emphasis the importance of data and a data transport network in current manufacturing. The interconnection between manufacturing and support groups will most likely be a LAN, a data switch, a PBX, or a combination switch that will provide access to the information needed to coordinate the office and manufacturing functions.

Various forms of information are communicated within the office, including work processing, electronic mail, document duplication, information storage and retrieval, conferencing, training, and fund transfer. All of these services are provided today in some form. Based on present experience, the secret to a successful operation of the office of the future is the way in which these services are integrated and the way people are trained to use them.

Organizations must be aware of the changes going on and must prepare themselves for a different type of operation in the near future when these changes can be implemented. Not only will these innovations change the way a corporation communicates, but they will also change the marketplace in which the corporation must operate.

A major decision facing someone in the telecommunication industry is whether a supernetwork will service all traffic or whether separate networks will emerge for voice, data, or video with interfaces between them. To address this dilemma, we need to review the nature of the network traffic, service criteria for these networks, and the homogeneity of these networks. The questions of security and "putting all your eggs in one basket" may also be critical to the decision.

It is necessary to subdivide the call types and their traffic into the different possibilities, from a networking standpoint, and then decide whether there is sufficient commonality to warrant merging from an application, managerial, or service criteria viewpoint. The combining of networks, from a traffic standpoint, will yield vast cost savings from efficiency of media, noncoincidence of busy hours, or many other traffic reasons. Offsetting these savings will be penalties for increased bandwidth and special facilities. Most of the traffic savings are associated with monthly charges, whereas the penalties will result in a higher initial cost. The economic selection, therefore, must be based on those costs incurred over the life of the network, and not just initial cost.

One obvious advantage of a single network is that the framework will include all the communications and processing functions. This means minimal end-user involvement and high productivity of application and system work. The disadvantage to a large complex network is its inability to add new features and services.

An analogy to this drive for a combined network can be drawn from the transportation industry reluctance to merge services from the trucking, railroad, and airplane industries. The argument of these industries was that the transport system used by each was sufficiently different to justify separate systems. The transport system for telecommunication is bandwidth, and whether the same argument will prevail for this industry is yet to be determined.

Although transportation companies claim they all have different criteria, their basic function is to transport goods and people between destinations: hence they are homogeneous services. Recently, letter/package carriers have incorporated many of the attributes needed for portal-to-portal transportation services. The same is true for telecommunication: If its basic function is to transport information without regard to requirements, then a combined network will prevail. If the service criteria for the different calls and messages are the driving forces, then the concept of separate networks will dominate, and telecommunication will function in a similar manner to the transportation industry.

It is extremely important that this distinction be made during the planning phase of the network. The planning of the network must be made with the thoughts that obsolescence can be avoided and services can and should be combined.

Another important consideration is that the planning cannot be accomplished with an ad hoc committee or a one-time plan. The technology and the services are changing so rapidly within telecommunication that it requires full-time attention to ensure that the services and systems are up to date. We cannot rely on vendors to keep us current, although they are a great source of information. We may want to avoid employing half-a-dozen people to stay abreast of telecommunication technology and use consultants instead. Our job is to implement a network and keep the users happy.

The planning of the network must have some definitive goals other than "modernize our processing equipment," or "get us good switching by the year 2000." More definitive goals are

- Make the network 100% digital
- Allow the users to have one instrument for voice and data
- Reduce the telecommunication long-distance cost by 20%
- Establish ownership without degradation of service
- Provide desk-to-desk calling without an attendant

In addition, there will be goals related to user improvements expected from the network, although some of them may be in the form of productivity improvements and be difficult to measure.

Most PBXs currently being purchased are for private ownership, and the next step appears to be ownership of a network that would provide corporate internal communication regardless of the location and still provide access to and from the public network. Deregulation and the demands for networks to provide more than voice communication are pushing business to investigate ownership of networks. Ownership of the network may be a prerequisite to survival.

One additional item on planning: The network we plan will be out of date when it is implemented. But that does not mean it is not cost-effective and providing excellent service to its customers. It is somewhat analogous to buying a personal computer; one year later the most recent model is able to do more for less. We don't throw away the computer because something new is on the market. The same is true for a communications network. As long as it is cost-effective, capable of providing good service, and can implement new features, we stay with the system.

Many organizations have a committee that oversees the data processing function and reports to an executive group for final approval. This reflects the large share of the budget devoted to data processing. This committee, consisting of part-time and full-time members, has the total responsibility for the data processing function. The same serious consideration must be given to the telecommunication function. Many people see this effort as an extension of the committee presently overseeing data processing activities. If this approach is taken, then the committee must include experts on switching and transport media as well as data processing.

11.2 Major Network Services

The principal reasons for employing a private network are the economics available from the ownership and the ability to combine services to lower cost and improve productivity. A quick review of these services is presented here.

The first, and probably main, service for the network is voice. Although we may want to say "elementary," many data processing people constructing networks today do not understand that voice is still the principal ingredient. Voice transmission is changing as a digital bit stream replaces an analog input. Although there is a lot of talk of transmitting voice in the digital format at lower than the 64-kb/s rate, a current network plan should only consider this rate. The network is converting to digital, and this evolution from analog to digital necessitates a strong standard; the 64 kb/s provides this.

For data the dominant factor appears to be the transfer of information and access to data bases. The speeds and quantity of the data devices determine whether the data should be combined with voice on the network (in many cases modems are needed where analog facilities are involved) or whether a separate data network is necessary. The separate data network can be a LAN for local solutions or a packet-switching network for long-distance solutions. In either case the plan must include an eventual merging or interfacing with the voice traffic.

The third source of information, image or video, is where the technology is changing most rapidly, with no clear indication of how it will finally be resolved. The X.25 and other protocols did wonders for data networks, and video needs the same considerations. Caution probably is the best way to address video, although the economics might favor some type of video arrangement over airline tickets. If a method of switching between data and video or a leasing of video facilities can be easily implemented, this should be seriously considered until standards are associated with video switching.

In addition to these basic services required with a private network, unique services are being introduced into telephony that were not there several years ago, and they signal the emergence of a market-driven force within the business. Among these services are:

Remote call forwarding   Callers can reach distant businesses by dialing a local (normally seven-digit) number. The call is forwarded to the business's central location, thereby eliminating the requirement for certain branch offices to handle customers' inquiries or complaints. This service is normally available in a variety of "packages," depending on the needs of the business.

Satellite service   The ability to move large amounts of information across the country or overseas rapidly. Several offerings are available with satellite service, including 1.5 Mb/s, television, or audio along with arrangements for control and use of the earth station.

Analog service   Opportunities still exist for using analog service where a point-to-point medium-speed circuit (9600 b/s) is needed. As digital switches start to dominate the market and digital facilities are employed, PBXs will start to use pools of modems for access to these devices. Thus all lines will now be able to use the analog facility without requiring a telephone with a modem attached or built in.

These services are only a few of those available and, more importantly, are now becoming price competitive. This is an important aspect of these services, because previously the cost was prohibitive, and only special applications or federal agencies could afford them. Now, with the cost being reduced through technology, the services are more readily available through new facilities and the competitive aspects of the current environment.

As networks are purchased, services, such as access to airline reservation systems, which were cost prohibitive when a connection was required from each office over an expensive facility, can be revisited to determine whether a connection from the network for all offices can be justified. In addition to the economy of size, the cost of the facility should also be checked, because the price is probably lower than it was several years ago. In other words, the decision made before the network was in place may change when these options are reviewed.

11.3 Network Standards

The network planner must deal with the standards and criteria needed for the network. Standards can strike down the finest of plans, especially in telecommunication, with its different standards for transmission, switching, power, environment, and terminals. Few people understand the standards for one of these categories, much less all of them. To overcome this problem, I have some great advice: Use the standards established by the Bell System for telecommunication categories. First, they have had over 100 years of experience, and you would be foolish not to take advantage of that. Second, the Bell System has performed an excellent job in meeting standards, yet returning excellent profits. Let's face it, that's all we want to do.

Some networks have been built on the premise that the standards set by the Bell System were too high and economic advantage could be gained by lowering these figures. Not so. The Bell System standards are set at good service, and, as it turns out, anything below those standards gives poor service. The reason is that probabilities have a way of ganging up on discrete events and causing a disaster. For example, a guru from Bell Labs/AT&T once told me that the fiasco in New York (poor telephone service) during 1969 was caused by New York Telephone Co. placing the standard for incoming calls to switching systems at P.02 (two calls in a hundred would be lost during the busy hour) instead of P.01 (one call in a hundred lost). This slight change apparently has a cascading effect on the service throughout the New York area, resulting in a tarnished image not only for New York Telephone Co. but for the whole Bell System (a cover story in *Time*). The ironic thing is that the cost differential between these two grades of service is so small as to be imperceptible on the balance sheet.

Other examples: An earthquake in California or a student on a tower in Texas shooting at people can bring the total long-distance network to its knees, although the occurrences are not nationwide.

The Bell System, I believe, has achieved the greatest balance between service and economics for telecommunication. I'm sure it is to our advantage to profit from this 100 years of experience and adopt their standards. It also makes thing a lot easier.

11.4 Network Design Considerations

The provision of sufficient bandwidth and facilities to handle data and voice traffic is the most important consideration in the design of a network. However, there may be many other characteristics to consider. For example, if 99.5% of the traffic is voice and the data traffic is slow-speed, then the network only requires a modem and a way to evolve the network from analog to digital.

The traffic pattern or flow of information is another critical aspect of the network that should be optimized. You may want to have a less-than-optimal routing pattern during the years of transition from analog to digital when separate facilities are required. This is only necessary if some type of digital overlay network is employed with the network.

These overlay services, like other telecommunication services, must be carefully analyzed from an economic viewpoint. Almost all telecommunication decisions are based on the economics of sharing: the greater the sharing, the greater the savings. The present telecommunication network is based on providing the callers with a continuous circuit during the connection. An exception is TASI (time assignment speech interpolation), but it reflects only a small portion of the network. For economic reasons the occupancy on these continuous circuits must be as high as possible, and most linking arrangements between offices are determined by usage and economic linking studies.

MERS

One example is high-usage linking between two points where a portion of the traffic is carried on the high-usage group, and the rest (overflow) is routed via a tandem or equivalent office and combined with other traffic. This overflow varies with link groups and depends on the cost of a high-usage link, the cost of the links on the alternate route, and the cost of switching the call through another office.

The industry for years only had high-usage linking available as a means to prove partial routes, but now there is most economic route selection (MERS), which can provide many more economic options in routing a call through the network. MERS may be the most user-oriented feature ever offered in a network. A comparison between alternate routing and MERS is shown in Figure 11.1.

Fig. 11.1. High Usage and MERS Routing

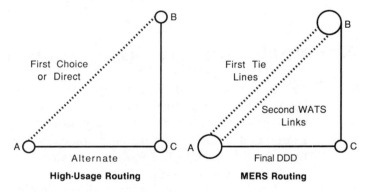

This comparison has been made previously and will not be repeated. However, it is worth repeating that MERS capabilities may be the most critical feature within a system. Many vendors are short-cutting this area to save memory. In taking this approach, the user can spend substantially more for long-distance service.

Other Routing Options

There are other routing options that the customer should explore before deciding on the final plan. Ultimately the system may select the cheapest route at the proper bandwidth for the call, but the dynamic of this is still years away. Meanwhile, the present selection processes have options available to help pick the best arrangement.

One such option is to transmit the call through the network as close to the destination as possible and then pass it to a commercial carrier for completion. This will work for voice calls, but there are problems handling data calls or calls requiring special bandwidth. From a customer viewpoint this is probably the cheapest arrangement, depending on the capacity of the private network. If the network was sized to only handle intracompany calls, a congestion problem is possible when additional traffic is handled. Careful inspection of the traffic must be performed before this service is implemented. It may be determined that at certain times of the day (nonbusy hour) this is feasible, and investigation must then be launched as to whether the switching systems can route one way during the busy hour and another way during the nonbusy hour.

Another scheme would be to have the translation portion of the switching system select the most economical route. This selection must be flexible enough to change the routing at certain times of the day to take advantage of variable rates offered by the carriers. Within most systems' translation, this requires tremendous memory and may not be practical with current technology. However, a routing node accessed over a special signaling link (SS#7) could provide a data-base resource for this type of information. This information is then passed to the translation of the system, and the call is routed according to the least-cost routing scheme. The translation of the switching system would require a default routing in case the link to the node was down. This node can be within the private network or, providing we have the proper routing interface, located outside the network. The latter would be the preferred method because updates to this information would be a time-consuming effort.

A third and less dynamic method is to establish a program where the translations of the switching systems are updated periodically to reflect the recent changes to the prices. This method requires access to cost information from the carriers and a program that would indicate whether any changes to these costs impact the network. The program would print which systems and translations are changed.

11.5 Network Routing Structures

Voice Networks

The hierarchical structure has been employed within telecommunication for years, but is now starting to fade due to deregulation and the introduction of LATAs (local access transport areas) to the network. Hierarchical structure is shown in Figure 2.4. Within this structure a call could go directly to the destination office, route via tandems within the chain of the terminating office, or route to tandem offices within its chain. To prevent routing in circles, a call is routed toward the destination; that is, the call was routed up its chain or down the chain of the terminating office.

Modern routing structures are represented mainly by some type of grid pattern, typically a two-tier grid. This arrangement allows the network to take advantage of the various linking arrangements and still retain the overall service criteria of the hierarchical configuration. A two-tier grid is shown in Figure 11.2.

Fig. 11.2. A Two-Tier Grid Routing

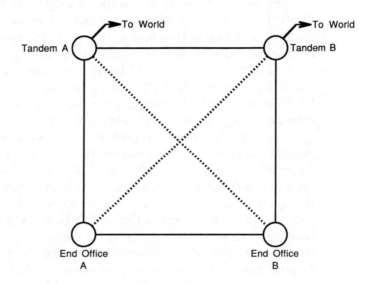

The routes to the world represent the various groups to the interconnect carriers. LATAs are finding this two-tier structure efficient for their routing problems. It allows them to combine traffic from several offices to one interconnect carrier. With the vast number of common carriers, this is almost a necessity. For states where there are several LATAs, this arrangement can provide a point for interLATA linking if or when it is permitted. The arrangement also permits one tandem to route all traffic to/from one sector of the network (country) while the other tandem interfaces to the other sector of the country.

Data Networks

Most data networks are arranged in a grid configuration quite like the two-tier grid for voice. The linking between these switches normally is configured to be fully interconnected, whereas the two-tier grid may not be. A typical data network is shown in Figure 11.3.

Fig. 11.3. Data Network Routing

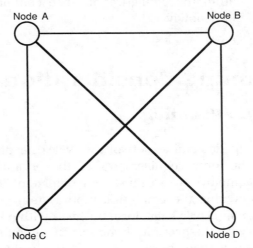

The advantage of the fully interconnected arrangement is that the failure of one node will still permit the routing of traffic between the other nodes on the network. As the network grows, however, the economic justification of maintaining links between every node will be difficult, and a modified two-tier grid probably will start to emerge.

This expansion of the data network and the emergence of the digital voice network (or nearly digital) as a two-tier arrangement will necessitate the economic evaluation of combining the tandem switches for both configurations.

The combining of tandem switches is achievable in today's environment because both networks are switching T1 at the tandem level. Also, data networks have modified the "random" packet switching philosophy formerly used. Most connections are being routed through data networks in what is known as a *virtual circuit* connection. That is, the connection between two users on the network is given a certain routing determined by the traffic load on the network at the time the connection is made. This routing is utilized for each packet unless an unusual condition occurs that would necessitate a node employing a different routing for a packet. If the original routing is used for the entire call, then the packets are received in the same order as they are sent. The overhead associated with rearranging the packet is greatly reduced with this scheme. It also means that packet switching has taken a giant step toward circuit switching. The possibility of these two call types being combined within the same switch—especially a tandem switch—is greatly enhanced with this arrangement.

The planning of the network should include a configuration where the voice and data traffic are combined at least at the tandem points within the configuration. The combining at the end office will evolve more slowly.

11.6 Planning Considerations

Traffic Planning

To determine network services needed to satisfy not only the various business requirements of the organization but also the needs of the employees, we must first determine whether there is an advantage to owning the communication system or staying with the telephone supplier. For a small local business with no desire to expand, it is best to stick with the present phone service. If communication requirements are minimal, then there is no need to consider a special system: the telephone company will provide adequate service at reasonable cost, and the small savings would not be worth the effort.

For planning a network, new talents are needed, many of which reside in the traffic area and include

- Traffic network design, which includes the switch location, the routing, and the trunk group sizing
- Analysis of various types of traffic data
- Determining the administration methods to use
- Network management techniques
- Analysis of switching and signaling problems
- Studying alternate routing

This list is not intended to frighten anyone, merely to warn everyone that it is not an easy task, but I can assure you that the rewards, both personal and financial, are very gratifying. Each area will be explained, and some type of methodology will be provided so that we can at least estimate our needs.

The traffic network design is concerned with placement of switch(es). If several locations are involved, a review of the requirements at all the locations must be performed before a final decision is made. Obviously, if the organization is served by a large headquarters located in the middle of the country, that is the place to put the switch. Decisions are not always that easy, and it is more likely that headquarters is on one of the coasts and the rest of the business is spread throughout the country.

Some businesses must control their phone system (e.g., catalog houses). These businesses should be analyzing their telecommunication needs even if they don't buy a phone system. If the business is totally dependent on the phone system, a constant audit of the cost and services must be done. It should include an analysis of the link groups to other areas and the cost of routing calls through the network.

Any analysis of telecommunication service must begin with a study of the traffic on the various routes and through the switches. This type of traffic study has become, to many people, a black art. Much of this mystery has been caused by the expressions used by teletraffic individuals, including erlang, poisson, CCS, BHBD, and so on. However, even a neophyte can appreciate the fact that a higher occupancy on a group of links can provide an economic gain.

The determination of the line traffic, both voice and data, has been discussed previously. It is important to size each switch properly for our own knowledge and for preparing a request to the vendors. Figure 11.4 illustrates the traffic flow in a switch from the viewpoint of determining traffic through the switch.

Fig. 11.4. Traffic Flow in a Switch

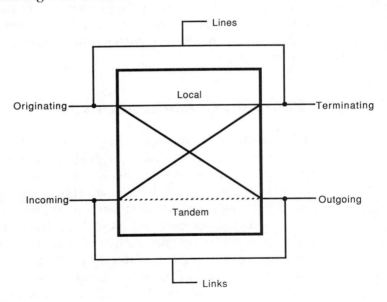

Regardless of the configuration of the switch, the traffic flow is from originating plus incoming to terminating and outgoing. The line traffic consists of originating and terminating traffic, which, unless you know otherwise, can be split equally between originating and terminating. We should be able to obtain the incoming traffic from the phone company, or the present switch may be able to provide that information. In either case this information is necessary to the study. Incoming traffic is normally destined for the lines (terminating traffic) unless the switch is used as a tandem or toll switch. If toll or tandem, the dashed line shows incoming to outgoing, and this figure must be determined or derived from the other information.

This chart can be used for sizing the criteria factors associated with the switch. Place the traffic figure for the lines and links at the originating, incoming, outgoing, and terminating points on the chart. The lines will generate originating and terminating traffic and, as previously stated, the traffic should be split between the two. For two-way links the same split should be assumed unless additional information is available. For one-way links their traffic will only appear on one side of the chart.

On another copy, place the originating and incoming calls to the switch, and the vendor can use this information for sizing the processor. On a third copy place the number of lines and links to size the switch. The vendor can add any terminations needed by the architecture of the switch together with their factor for miscellaneous items requiring a connection on the network. Indicate the percent growth anticipated per year, and this should give sufficient information to obtain a quote from several vendors.

Other Planning Considerations

The centralized maintenance concept allows the network to be operated at a minimum cost because individuals are not on site around the clock. To the owner of the system this is also the most visible indication of the network and how it is performing. The vendor must be able to explain the various night options associated with each switch.

The digital interface to 1.544 Mb/s is the way to get to the backbone of the digital switching hierarchy presently being introduced. This is also a critical feature and the method by which it is introduced between a system and the rest of the world is equally important. The vendor should be able to demonstrate their ability to interface their system with a digital stream. Two major options should be demonstrated: one related to the vendor running the system from the master clock, and one where the system is synchronized from a Bell carrier group. Depending on the application, we may want one or both of these options. The preferred method is to have the system sync from a Bell unit, thereby ensuring compatibility with the network. The Bell System has a very attractive rate on T1 lines.

Common-channel signaling is another prerequisite for a proper system network, because it has access to services in addition to optimal routing. This includes special data bases, special routing information, unique networks, and, most likely, access to the main long-distance network. An additional advantage of common-channel signaling is its ability to provide simple link access between the offices within the network. However, several systems are available, so care must be taken in obtaining one that will provide the interfaces needed for the network. Common-channel signaling can provide the means whereby the user has control of the network when placing a call. Presently, this function is not available with common-channel signaling, but it can be added.

An access to a common data base will provide the system with a great deal of growth capability. This capability will be the method whereby the system can add features without requiring a major restructuring of the system software. Today's systems require as much time to add new features as it took to develop the original software. Many of these new features will be capable of being placed in a common data base, thereby avoiding the expensive development. These data bases would include things like special translations, special directories, unique services, and routings.

It is expected that the ISDN will include the ability to have a digital phone interface to the network. Many vendors are looking at the digital instrument as a way to have voice and data units at the customer's desk without requiring two pairs of cables to the switching system. Therefore, we can expect a proliferation of digital phones to enter the market within the next few years. The digital rate can vary from 132 to 256 kb/s, depending on the vendor. The basic arrangement should ensure that 64 kb/s is transmitted from the instrument to the system and 64 kb/s from the system to the instrument, plus a 16 kb/s control channel.

The basic transmission rate of 64 kb/s is important because this is the North American standard.

The interface to the packet network is extremely useful for the occasional use of a remote computer facility. The interface requires the system be adaptable to the X.25, X.75, and other standards necessary to make use of these networks.

Once the network is purchased, the administrative methods can be critical if the service is inferior to that presently provided by the local phone company. The administration, in addition to being concerned with service, must investigate growth (up and down) opportunities to ensure optimal return on investment.

One major problem of the administration is who should be contacted when the system exhibits a major switching or signaling problem. No more calling the local phone company for help. Many companies opt for a maintenance contract with their favorite vendor to resolve this problem. Many vendors will offer a guarantee with switching equipment and transmission gear for a time, and the contract should not be let until after this period.

11.7 Private Telecommunication Networks

Requirements

The new generation of private networks will require a vast array of features and services in addition to the basic voice and data functions presently being performed. These features include sophisticated call routing, special billing, centralized maintenance and data bases, and access to an SS#7 link. The various types of PBXs in the network were built to handle voice traffic, although many of them are based on digital switching concepts. Most of the network facilities are analog or leased from Bell. The question is how to move from the present environment to a full feature private network arrangement. Some examples are shown, illustrating possible configurations, but first a discussion of the most critical element of the network, the PBX. PBX is a somewhat generic term and, when used, will also refer to centrex or any switch attempting to enter this market.

Essential PBX Features

If a private network is to be realized, then each PBX must have certain essential features needed by corporations. Many corporations depend on the collection and dissemination of information, and the PBX features must cater to this need through various accesses and transportation media. Many of these features are different than the user's features normally described for PBXs. Among these are

- Remote or centralized maintenance

- Digital interface to a 1.544-Mb/s digital stream
- Some method of common-channel signaling
- Access to common data base
- Interface to LAN
- Line interface up to 64 kb/s
- Packet-switching access
- Centralized attendant
- Billing functions
- ISDN access
- Network and routing control

Some additional features that are highly desirable with the operation such as a common feature set to allow users at all locations to learn the use of a particular feature without being retrained when they move to a different location. Also desirable is a common traffic data output package, although the network control center can sometimes reformat this information into common output.

Several architecture structures will satisfy these features, and most vendors are working on arrangements for this specification. A problem occurs for most vendors in deciding whether the access to a feature should be an integrated part of the system architecture or an adjunct to the system. For example, the access to a packet network (the same principle applies to a LAN) is best handled within a switch, whereas access to the maintenance center can be an adjunct to the switch to accommodate the unique interface.

Many of the features are known implicitly by users of telecommunication, but vendors have a way of focusing this understanding into the way their system handles calls. A few of these features need to be explicitly stated to develop a common base for the requirements.

Routing

Most modern PBXs have some type of MERS. In most applications the routing follows the pattern shown in Figure 11.5. Note that the computer moves from the least expensive to the most expensive in an orderly manner, and this routing is transparent to the customer. The administrator requires information on the amount of overflow from the routes together with information on the traffic carried on each route. This information can be used to decide the optimum size for these groups, because overflow traffic does not act the same as first-routed traffic, and a complicated analysis is necessary to sort it all out (even I think the study of overflow traffic is disgustingly complicated).

Fig. 11.5. MERS Selection Process

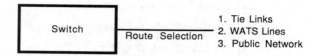

Another MERS feature required within a network is queuing for facilities. That is, certain customers may find it more economical to wait for a facility to become available rather then route advance or return busy. There are two types of call queuing:

Off-hook   Wait for an available link

Callback   System will alert when link is idle

Off-hook queuing is normally used when the link group is of sufficient size to minimize the wait. In other words, you wait with the phone to your ear for an idle link. The callback feature allows the customer to hang up (go "on-hook") and have the computer call when a link is available. This feature is best with small link groups because the waiting time could be excessive, especially where long holding times are involved. The number of customers queuing for a group should be one or two to avoid long delays, and users should not be participating in more than one queue at any time.

These options are available today, and no system should be purchased without the equivalent of these arrangements. MERS is indeed the greatest user-oriented feature within a PBX (today), and if any vendor is reducing memory in this area to add other features, ignore them. If anything, a vendor should be capable of explaining how this feature can be expanded. Some future options for routing consist of

- Routing the selection for bandwidth with the least expensive being selected first

- Routing the call through the network as close to the destination as possible before going off-net

- Allowing dynamic routing changes during the day corresponding to the cheapest variable rate available

- Allowing updating from a remote location when rates are changed

Network Access

A substantial change has occurred to the long-distance network from the days of "all calls" via the Bell network. However, the selection of a carrier on price alone could be a tragic mistake, because we still want the voice quality we are accustomed to. Consequently, before selecting a carrier, we should test their service, either via another user or a trial arrangement with the carrier. Then we sample the various users to determine whether the service (transmission, wrong numbers, etc.) is equal to or greater than the present service. We don't need to settle for less; there are enough carriers around to supply good service.

For a private network a number of technical items must be addressed, which can serve as an aid to developing a checklist for evaluation. Assuming there is a position on the long-distance carriers, additional consideration would be billing, operator assistance, common-channel signaling, and grade-of-service selection.

Billing and collection are sometimes tedious tasks, depending on the amount of detail required to operate the business. If the business chooses to handle long-distance traffic itself or adopt a bypass strategy, call billing is its responsibility. Programs can be obtained that will provide billing data, and the data processing system can sort the information in various ways, depending on the requirements. However, the problem is that the majority of PBXs interface to the public network over loop links (trunks), and, as such, no answer supervision is passed from the public network to the PBX as an indication when billing should start. Without this, you cannot accurately perform billing functions. This situation must be examined at each location where billing information is to be collected.

The signaling method over links between various locations is changing as the network moves from per-link signaling to common-channel signaling. This move will eventually allow for a clear data or voice link that will not be encumbered by the need to pass signaling or supervisory information. This signaling method is not only important when developing a communication system but is helpful when evaluating the carriers. A business may want to delay billing until some type of common signaling system is in place.

Other considerations in the area of access include grade of service, remote switch service, and overload control. The grade of service is related to the economics of the network. The Bell System assumes an end-to-end service level of P.04. This means that 4 out of 100 calls will fail due to blocking somewhere in the network (including switch blocking) without regard to the status of the called customer's line. Many time-division switches will provide nonblocking switching that allows the P.04 to be concentrated in the links between offices.

Many networks are offering remote switches to handle branch of-fices or off-campus areas; the service level at these locations should be excellent (P.001), because these switches are considered adjunct to the local switch, and the users should be obtaining the same service level as the local user.

Overload control is concerned with traffic overload and switch out-ages. Dynamic controls are the optimal way to deal with both conditions, but this might not be possible initially due to existing conditions within the network. If not, a plan must be incorporated to reach this goal.

Centralized Attendant

With INWATS or 800 numbers and a company network, a centralized group of attendants can be very attractive from an economical viewpoint. This service can be supplemented with local computerized directory assistance voice service.

Because many operations, especially department stores, also require local numbers, the centralized operator must be accessed over a link, during which time the operator will key the destination number after the required service has been determined. The link to the operator must then be released, and the calling customer connected to the proper service. This feature must work in a similar manner when a call is being transferred from department to department with the assistance of the operator. Without this "releasing" of the link, the network would quickly become congested.

Data Access

Everyone knows about the numerous features associated with voice traf-fic on any PBX, but little is known about the data features these switches can or should be offering. Several operate in much the same manner as the equivalent voice features: message storage, paging, and conference calling are a few examples. Data may only occupy a small percentage of the switch traffic, but can add significantly to the requirements of the switch.

The major issue for voice/data is the subject of a single integrated path for the customer. That is, the terminal at the customer's desk is capable of converting voice to a data stream and of transmitting data over the same pair of wires. When exploring features for data, we must avoid denigrating the voice connection. To accomplish this balance with all the features intact requires a machine that can handle the conflict between these services. For example:

A data call cannot be interrupted by the various tones (for example, call waiting tone) presently being used on the voice network. However, most data users will want an indication that a voice call is present. How this dilemma is solved by the PBX is worth investigating.

The terminal must also be capable of recognizing whether the incoming call to an idle line is a data or voice call to prevent a computer from calling a voice-only line.

The most economical routing feature of most PBXs can be extended to data call, but bandwidth should be included when laying out this routing. Also, stations may have different data speeds, and this identity must be included.

Billing of data calls is as important as voice calls, although the approach may be different because a data call can be via a packet switch, and the user should only pay for time when packets are being transmitted.

Access to and control of information is extremely important in today's environment. The business world is becoming inundated with terminals connected to something. The most visible part of these terminals is the input/output (I/O) portion. The standard input today is a keyboard that looks like a normal typewriter and is basically used the same way. Each keystroke is captured as a digitally coded signal used by the terminal for transmittal to the data processing host.

Output to the terminal or user is accomplished in several ways: hardcopy from a printer, softcopy displayed on the screen, or both, depending on the requirements.

The terminal or DTE can use the 3-kHz telephone channel for data rates of 1200, 2400, 4800, or 9600 b/s. The speed is determined by the DCE, and this unit is normally separate from the DTE for more flexible arrangements. The interface is defined in EIA Standard RS-232 or CCITT Recommendation v.24 for the 25-pin plug and socket. For 9600 b/s the line must be conditioned—that is, no bridge taps or loading coils—and have private access.

Line Interface

Once the network has the ability to connect both voice and data connections to the customer loop or to the PBX line, it is the responsibility of the switching system to separate the functions and pass them to the appropriate network. A voice-switching system and a data-switching system may appear similar, but they are quite different. Any attempt to incorporate both functions within one switching array has resulted in disaster or great difficulty. However, experience helps, and the current switches are beginning to show some ability to handle economically both types of traffic.

The current approach to switching an integrated workstation is to employ different bus structures, one for data and one for voice, with the line equipment being connected to both buses. One example is shown in Figure 11.6.

Fig. 11.6. Line Interface

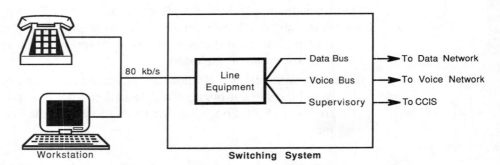

The current switching systems are hard pressed to provide these functions and remain competitive. However, the customer's demands and the lowering of IC costs will open the door to these functions within a system. The billing, maintenance, and administrative features of these calls are distinct and, hopefully, will lead to separate nodes. Complexity of software is a major problem not only during the initial development phase but also for updates. It appears that functions eventually must be separated to ensure faster introductory cycles and simpler systems.

The subscriber presently has access to the network via a single pair of copper wires, and though it would help if instantaneously all subscriber lines were replaced with an optical fiber link, this is not about to happen. In the interim various schemes are being worked out to transport voice and data over the present wire pair. One such scheme is the use of 80 kb/s between the customer and the serving office. The 80 kb/s breaks down to

- 16 kb/s data
- 16 kb/s signaling and supervision
- 64 kb/s voice

The time separation between transmit and receive is performed by time compression or bit interleaving and colloquially referred to as *ping-pong*. With most current arrangements the data terminal is an extension of the phone, although ultimately we will have a workstation with integrated data, voice, and other information services. Many applications can be satisfied with the ping-pong approach and should be a viable solution during the network transition. The major advantage is the ability to have a voice call and a data connection simultaneously.

The workstation or customer subset will consist of two parts: the customer equipment and a network termination. The network termination will have the responsibility for providing the proper interface and for addressing the network with information on what services are being requested. The network termination, via this interface, can be aware of the services possible from this customer and will route accordingly. Some examples of these services are

- Voice only
- Data only
- Special electronic filing
- Data entry
- Combination voice/data
- Combination voice/data/electronic mail
- Video text
- Word processing

The basic aim of these services is the increased productivity of the user by offering more conveniences and problem-solving abilities. Part of this convenience is one access for all services, the goal of the ISDN.

12

Network Illustrations

So now we have come to illustrations, which seem to be a different approach. We have tried to equate networks with some property or properties that can be related to organizational goals. Illustrations pronounce judgment of a different sort: This is how we see the world, not as features, standards, and services, but the total interconnection, a miniuniverse. No illustration can completely capture another miniuniverse; it can only serve to stress some important points.

12.1 Objective

The objective of the network plan is to determine the optimal evolution of telecommunication with a particular type of equipment and minimum cost. The evolution of the network means expansion or growth of the equipment will be possible and cost effective. A guideline to the latter is "a network whose cost increases linearly with size."

If the network consists of a few PBXs that can be tied together through judicious use of codes and levels, I don't think there is much need for total network planning. However, if the need is for a private telecommunication net capable of voice, data, facsimile, and video plus some special features, network planning is called for.

The information on the network should initially consist of a list of the offices within the network and whether they are used as tandem or transit offices. Information should also be obtained on the overall routing of the network and how each office interconnects with other nodes on the network. This is called the *trunking diagram* for the network. After it is constructed, decisions can be made on whether the routing and office placement should remain the same or be redone.

A critical factor in the routing structure is the present rate for interconnecting various offices or leasing facilities from AT&T or other common carriers. For now, we assume the network can be optimized from a traffic standpoint.

The office placement is a vital part of any network, and consideration must be given to a program that helps to do this. Throughout the evolution of switching networks, researchers have been attempting to devise ways to switch more traffic without major changes to the system. One prime example was the creation of the scheme to assign a channel only when someone is speaking, thereby doubling the number of conversations that a group of links can carry. The placement of data on the section of the bandwidth that voice connections are occupying but not using is another successful scheme. The basic principle to remember is that an in-place network quite frequently can yield substantial improvements without a vast outlay of money.

12.2 System Sizing and Placement

The easiest way to determine where the principal switching systems should be located is to already have switching at very large locations. In this case we need only replace the Bell System gear with our equipment, add data and other traffic to the voice requirements, which are known, and determine the facilities to handle the increase in traffic. Easy. Unfortunately, most situations will be much more difficult than that, including determining the various locations for switching systems.

One method to make the decision easier is to locate the various sales offices and manufacturing sites and interconnect them with headquarters. After this, the common carrier routes can be located, unless the business wants to own its facilities. Several common carrier locations should be established. The selection of major locations for switching systems should be achievable now by either observation or careful analysis depending on the available tools.

Arrangements must be worked out with the data processing personnel in sizing the various data requirements. These requirements include facilities to and from the switching systems and the facilities interconnecting the various locations. If the local switching system must interconnect the LANs with the mainframe or other locations, this traffic flow must be sized and the interface problems resolved.

Initially, a common carrier can be used for major routes and facilities. This is a good idea even if the company facilities could prove in; because, first, the business has enough things to do and this can be turned over to the common carrier, and, second, the business is more concerned with the type of facility rather than the ownership. It can later own the facilities, although fiber optics may make leasing more attractive.

If a common carrier is selected for the long-distance facilities, the planning of the interconnection will have to be in concert with them. The common carrier will be able to obtain frequency clearance, for example, if microwave is selected. If the network is expected to be T1 lines, deal with a common carrier who can provide T1. This will avoid changing the configuration later.

Network Modeling

We now have enough information to model the network. We can do this by inspection, or we can develop a sophisticated model. It is best if both are used. The inspection will get us started without waiting for the model, whereas the model will allow us to fine-tune the network and determine the impact of new requirements, new switching systems, or new technology on the configurations.

PBX versus Centrex versus Bypass

The battle for private networks will be at the forefront of telecommunication as various vendors from different segments of the industry vie for this lucrative market. If the network can provide the features and services demanded by industry why bypass? The ISDN of the near future promises to be all things to all users and, if true, the current trend of bypass will be halted.

The other question of centrex versus PBX is somewhat more difficult because it possibly depends on whether LANs can effectively transmit voice within a local area. If LANs can effectively serve voice, then centrex will probably emerge as the leading interface between local service and long-distance service for all telecommunication needs. Otherwise, the PBX will continue to be the dominant communication switch in a corporation network.

12.3 Voice Network Configuration

Because the voice network is in place, many of the problems associated with it must evolve over time. For example, the principal goal of 100% digital may not be an economic reality due to the present investment in analog switches and facilities. The goal is still a fully digital network from the switching and transmission viewpoints. The basic network structure will start to take on the characteristics of the arrangements in Figure 12.1. The T1 facilities are dominant between the offices. The digital facilities can be interfaced to analog switches or supported by some existing analog, but the T1 lines should be the dominant factor in the interconnection of offices. The figure shows the arrangement for one cluster, and the size of the network will determine the number of clusters.

Fig. 12.1. Voice Network Arrangement

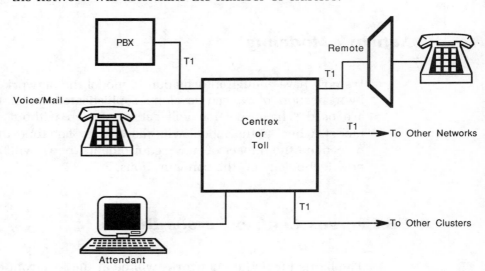

One feature of digital transmission is its ability to maintain a "pure" signal from the originating office to the terminating office. Noise, crosstalk, and loss are kept to a minimum. If an overlay network is used, the goal of a single conversion from analog to digital or from digital to analog (except at the station) should be established. Digital and analog switching have different specifications, and the quality of transmission can be affected by the constant conversion from one to the other. This goal will influence how offices and facilities are converted to digital and, in some cases, will override the economic plan for replacement.

Remember, although the overlay network is an intermediate goal, it is critical because the user's first impressions of the new network will come from this arrangement.

This approach probably will result in certain groups needing analog and digital facilities between clusters. The economic penalty for this will be small compared to the eventual transmission gain when the total network is digital. However, it will necessitate an investigation each time a switch or facility is converted from analog to digital. These groups should be used only where absolutely necessary so that constant analog-to-digital conversion is avoided. Features such as MERS will maintain the cost penalty at a minimum.

If we adopt the view to interconnect the data and voice information with the T1 facilities, then the problem will be to find a digital crossconnect scheme and a T1 multiplexer or the equivalent, which will satisfy the traffic and linking levels.

Between this network and other networks the use of T1 facilities is also recommended. Advantages to T1 for this access are the eventual combining of services over these facilities, the ability to maintain a digital signal across the total network, and economics. The latter will depend on the amount of traffic between the private and public networks. The proper selection of interconnection points between networks will, in many cases, yield sufficient economic advantages to T1 facilities.

12.4 Data Configurations

Until very recently, remarkably little was known about switching data traffic. There were no reliable answers to even the most obvious questions: How is bandwidth determined? Should the route be dedicated or switched? Can a terminal operating at one speed communicate with a unit operating at another speed? Can a PC communicate with a mainframe if both were purchased from the same manufacturer?

Happily, the experts of data and telephony have started to get together and determine methods whereby data can be switched and the amount of traffic can be estimated. For small data requirements a couple of modems and a private or public network are sufficient. As data traffic increases, an investigation into other data configurations must be undertaken. A number of factors will determine the final configuration for the network, with cost being a primary consideration. The cost factor varies widely with the equipment purchased and can change dramatically, due to the present environment. The examples shown here should be used as illustrations on how problems can be solved. The example intentionally ignores cost considerations, which should be one of the main criteria, because they are so volatile.

A particular corporation's data communication requirements are driven by a large number of terminals that need access to a mainframe computing resource located at the company headquarters. The basic speed for the terminals presently is 1200 b/s, but plans are to increase this to 9600 b/s. Most of the terminals are connected to headquarters over full-period circuits leased from the telephone company. The corporation is concerned about the high cost associated with controllers in front of the mainframe and the low occupancy on the circuits. This problem is raised to another level if, instead of terminals, PCs are located at some of the sites.

The interconnection of the terminals with the mainframe is complex. Some of the problems are transmission speed matching, code conversion, flow control, and delay times. The company wants the ability to interconnect with a data network and to share the facilities with the voice network. The data configuration is shown in Figure 12.2.

**Fig. 12.2.
Current Data
Arrangement**

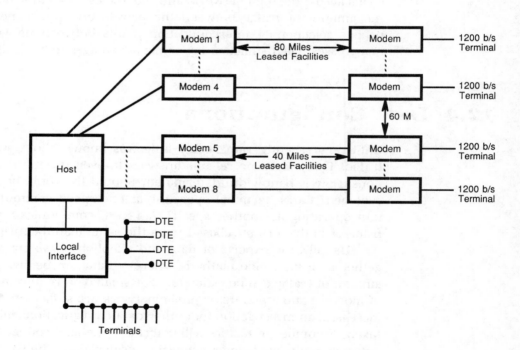

The data network is an elegant way to solve the data-switching problem, but it still requires several steps to reach that goal. The goal should include the use of the same T1 facilities for both voice and data, with the additional goal of accessing these facilities from the same switch and sharing them with video or other wideband needs.

The cost for this configuration can be found by using the cost for leased lines for the various distances and the cost for the modem. Usually the equipment is leased, and cost figures should be arranged to reflect this. It the modems are purchased, an equivalent rental can be derived or obtained locally.

One alternative is a multipoint connection combined with time-division multiplex to save line cost. See Figure 12.3.

This configuration allows us to take advantage of the latest technology and minimize the line cost, usually the most expensive item. This, again, should only be considered an interim step. In the final configuration the modems should be eliminated.

Although the arrangement provides some cost benefits against other configurations, the key consideration is the ability to have a facility that can handle voice, data, and video switching. The possibility should also exist for switching of the functions during special requirements or by time of day. For example, if a large portion of the T1 line is needed to exchange large volumes of batched data processing at night, arrangements can be made to increase the data traffic at that time. During the day the T1 facilities are employed for voice traffic and inquiry/response data traffic. Given the proper equipment. The T1 line also could be devoted to videoconferencing. This configuration is shown in Figure 12.4.

Once the T1 facility is in place, different arrangements and interim steps can be configured to minimize network cost and accelerate the prove-in period for the investment. The resultant network is a powerful and useful resource for the company, which will provide for economic and technological growth.

12.5 Example of Fiber Optics

Although considered the newest fad by many, fiber optics has been with us for quite a while. It began in the 1960s with experiments at Corning Glass and quickly proceeded to Bell Labs and others. It created quite a stir in the 1970s but quickly faded as a gimmicky diet when manufacturing and installation proved extremely difficult. The 1980s belong to fiber optics, because new techniques and implementation procedures are being marketed.

An example illustrates the widespread application for fiber optics, particularly in the telecommunication field. We give a hypothetical situation that can serve as a basis for value judgments on proposals that may

**Fig. 12.3. A
Proposed Data
Arrangement**

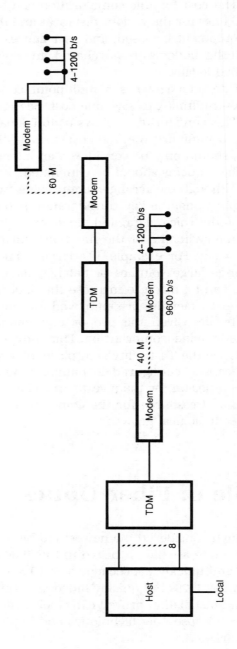

be encountered in the future. It will also show the myriad solutions for handling traffic between offices.

A requirement for six T1 systems between two offices has been established for a particular customer. This requirement may have arisen from a need to switch a number of voice circuits approaching 144 (6 × 24) or some combination of data/voice traffic reaching that level. This example will only deal with the basic connection problem. The distance problem can be solved by repeaters, and the growth problem can be addressed separately. It is well understood that distance and growth are factors in the final decision, but this example will concentrate on the basic connection and interfaces required to solve the problem.

Fig. 12.4. Data Configurations for Expanded Traffic

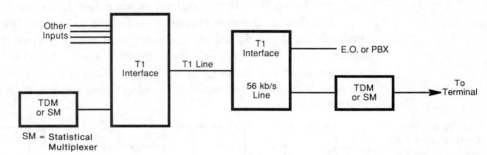

A radio link may be a viable solution to this problem, but for now a fiber-optic solution is assumed. One simple, straightforward solution to the problem with fiber optics would be to bring the T1 lines (1.544 Mb/s) to a DS1C signal level (3.152 Mb/s) through the use of three M1C multiplexers. See Figure 12.5.

Fig. 12.5. Fiber Optics with DS1

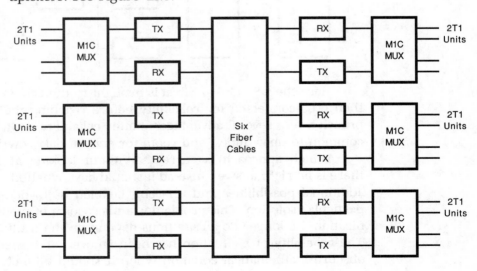

The TX and RX symbols refer to the transmit and receive legs of the unit, which are all four-wire circuits. A DS1C signal is applied to each fiber pair, thus allowing two T1 systems to be carried on each fiber pair. No protection is provided with this arrangement; that is, the loss of a fiber will result in the loss of one-third of the circuits. The aim of this example is to provide a simple solution to the problem and would more than adequately handle the situations. The traffic loss is based on the failure occurring during the busy hour, the worst case, but failure during other periods would not have as severe an impact on the service. Two factors should be examined when considering failures in a configuration: (1) the amount of degradation the failure will impose, and (2) the estimate of the traffic impact during times other than the busy hour.

Another and more elaborate solution to the problem would be to employ high-level digital multiplexing and protection switching in combination with wavelength-division multiplexing. Despite the long list of sophisticated equipment, this solution emerges as a viable alternative because it reduces the number of fibers and increases the reliability. See Figure 12.6.

Fig. 12.6. Fiber Optics with DS3 with Reliability

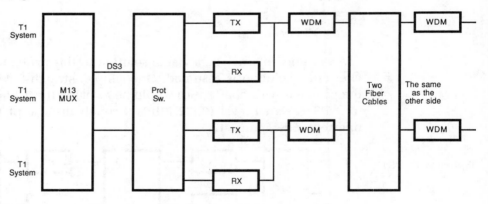

Here the DS1 (1.544 Mb/s) is brought to the DS3 (44.736 Mb/s) and then crossconnected for reliability before reaching the fibers. This approach offers several advantages: better reliability, switching at a speed common to fiber optics, and room for substantial growth if needed.

Which approach is correct? Without looking at the application, there is no right answer. Instead it should have whetted our appetites for additional possibilities and, thereby, enabled us to determine our own particular solution. This problem is a microcosm of modern telecommunication. No longer do all solutions depend on the traffic characteristics. The reliability of a situation has been ignored or buried beneath some obscure mathematical equation when it should be included in the customer's needs during the proposal stage. A sizable problem can be encountered during even a partial failure for those corporations completely dependent on telecommunication, and a complete failure is catastrophic. For these situations complete redundancy is required.

Thus, we might conclude that a broader, technical competence is necessary to deal with these problems. I don't agree. A person who understands what is going on and how dependent the company is on telecommunication can derive a solution as readily as the best technical mind.

12.6 Putting It Together

Now we have many options for implementing an integrated network. Each vendor says its way is best, and for a particular application the vendor is probably right. Units can be purchased that will connect a phone and terminal, allow simultaneous voice and data calls, and separate these functions at the PBX or switching office. If only a small requirement exists for combination voice/data, it could be the best answer. This section assumes we want to bridge voice and data systems to achieve an integrated network.

This section deals with the technical areas of a network, but this is only one part of a multifaceted problem. The current telecommunication environment requires patience for many authorization procedures. For example, if we need a microwave to transmit to a common carrier, it may take a year to get approval for the frequency. Harassment along the way from present users of frequencies in the area can be expected. However, a firm grasp of the technical aspects of networking will guide anyone through these times.

Figure 12.7 shows the configuration of a private network for the current voice and data arrangement. The computer center is equipped with leased broadband lines to various multiplexers. The multiplexers go over leased voice-grade lines to branch locations.

The T1 crossconnect schemes may have the ability to do several types of switching. The main purpose of these units would be to concentrate as much traffic, both voice and data, to T1 facilities traveling between the various locations. This technique is valuable whether the T1 lines are owned or leased. When T1 lines are used between the locations, an increase in traffic will yield an opportunity to move from DS1 level to DS3 level, at which the cost per call-unit will show substantial savings.

Each region is equipped with controllers or equivalent systems to serve the terminals and gain access to the main processing center. These

**Fig. 12.7.
Network
Configuration**

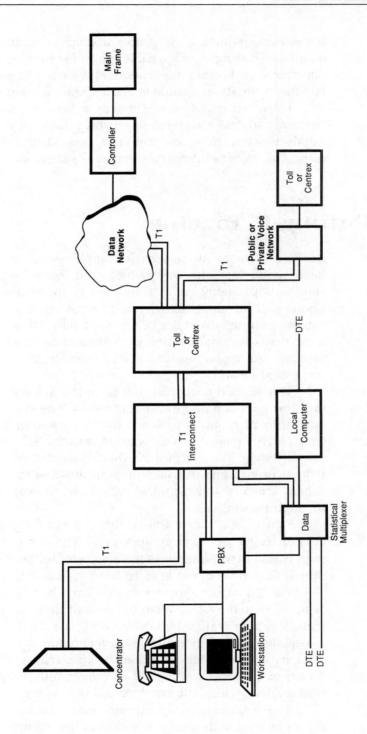

minicomputer systems are equipped with printers. Most terminals interface with the computers via multiplexers (controllers) that have a transmission speed of 56 kb/s. The maximum speed from a terminal is 9.6 kb/s.

The approach to the network configuration will be to arrange the data equipment into a packet network and the voice equipment into a digital network. The packet network will allow the terminals to be consolidated, the leased lines from the Bell System to be eliminated, and the LANs to be incorporated into the network. The digital voice network will provide a cost improvement over analog for transporting the call end to end. Both of these undertakings will be time consuming and delicate, because they must be accomplished without interrupting or degrading service from the user's viewpoint.

Once both networks are in place, a study on where they can be interconnected is included in the plan. The advantages will come from having groups of facilities between nodes rather than separate links for each network. Traffics associated with T1 facilities are becoming attractive and should continue in that direction as additional fiber-optic systems are installed.

The user, with a combination terminal, will insist on combining these functions.

Data Network

The data equipment can be brought to many configurations, depending on the traffic and network needs. The emphasis should be to use T1 lines for interconnecting the locations. The network will allow access from the user's terminal to many mainframe computers. The packet network, if used, should be based on X.25/X.75 to interconnect to other data networks.

Options for handling the data at each location are available. Which option is best depends on the equipment currently installed and the anticipated future of data within the network.

Data network   This arrangement employs switching nodes at each location to concentrate the data traffic to the T1 level and transmit it to other locations. If interconnection to other networks is considered, this arrangement is best.

SNA network   Equipment consists of IBM with FEPs at each location. In this case the multiplexers are used as concentrators to the IBM units. Best where a long-term commitment to IBM is anticipated.

Statistical multiplexer   The minimum investment, because these units can be reused when the networks are combined. The units should have the ability to select a computer or else connect to an existing IBM system that interconnects the computers. This title covers concentrators and multiplexers. But "statistical multiplexer" is used to denote a system with a processor associated with it.

The multiplexers can have functions that cater to a distributed network of terminals. A review of some of these functions is worthwhile.

Operate with multidrop lines   Where there are low-speed terminals attached to a line, the unit must be capable of selectively recognizing each terminal.

Buffering   For maximum speed for the network it is necessary to have some buffering associated with the various control units.

Support a wide variety of inputs   Asynchronous, synchronous, voice, facsimile, and control information.

Bypass operation   A T1 line can pass directly through a multiplexer on its way to a particular destination.

Drop and insert   The multiplexer can terminate certain T1 lines at its location and insert other T1 lines.

Bandwidth allocation   This is an optional feature in which bandwidth can be allocated during the day for particular functions such as video, file transfer, voice, or data. If there is to be a glut of bandwidth on the market as predicted, this feature will have limited value.

The various functions allow the network to operate at the least cost with excellent service to the users. Once the network has been determined from a functional standpoint, a review of the reliability needs of the system can be made. The distance between any two locations and the cost of the facility decides, in many cases, whether this duplicate interconnection is necessary. Initially, the cost may be prohibitive for all but the most critical locations, but it should be reviewed every year as prices decrease.

The functions at the various locations will be somewhat determined by the requirements of the customer. A banking network, for example, will need minicomputers to record and update records with only an occasional transmittal to the mainframe computers. An airline reservation system will have much more interaction with the mainframe computer.

Each configuration has to be priced against the present arrangement to take advantage of the savings in facilities and terminal charges. The selection of the proper plan is the second step, and it should be based on strategy rather than cost. The third, and most difficult, step is selling the plan to the organization. The fourth step is the implementation of the configuration. How well this phase goes depends on how well the third step was done.

Voice Network

The configuration for the voice network will parallel the data network; that is, it will consist of major switches with focal points at the main computer locations. Other locations will have digital PBXs or concentrators interconnected to the prime location by T1 facilities.

The switch at the prime locations can be a toll switch, a centrex, or a pure tandem, but, in either case, capable of handling tandem traffic without seriously impacting the line traffic. A centrex switch would be an ideal candidate for this switch if there is sufficient traffic capacity and current federal studies provide favorable rulings for it.

We can divide the functions of the voice plan into three categories: equipment selection, conversion from analog to digital, and feature determination. The objective of the equipment selection is to operate the various locations as uniformly as possible. That is, someone transferred from the East Coast to the West Coast must buy a new wardrobe, but he or she doesn't have to relearn the user's feature on the PBX. If a potpourri of PBXs presently exist in the network, either a planned replacement (especially if many are analog) is undertaken, or equipment is purchased to circumvent the differences. The establishment of uniform network features is important in the voice network but will be critical if the voice/data networks are combined. It is best to start the campaign now.

Facility

With both networks using T1 facilities, transmission between the location can be DS1 or higher depending on the traffic. Either microwave or fiber optics can offer cost benefits over many leased lines. Distance, terrain, right-of-way, frequency allocation, and other factors will decide which is used. The planner's primary concerns are cost of the links, the amount of voice traffic, the rate at which data can be sent, and available data options.

In addition to the voice and data traffic, low-speed transmissions are required in the network. The transmission of control and billing information from each location to the network control center is necessary. Some of this information is transmitted at low speed or late at night to minimize the impact on other traffic. Status or trouble information is transmitted concurrent with voice and data traffic and must be factored into the facility requirements. The status/alarm information for the voice and data facilities can be interlinked if the control center is common to both. This in itself may prove that once and for all an integrated network is cost effective.

12.7 Summary

Voice, data, and facilities are becoming interlinked for large networks. Any organization considering the incorporating of these functions in a network cannot advance piecemeal into the world of telecommunication and ISDN. A great deal of planning is needed before the first purchase is made. Also, some walls have to be torn down between the voice communication and the data processing groups. Both groups must operate from a single telecommunication base: the passage of information from place to place. The field of telecommunication is moving toward a unified network, and the new information-based individual will have a significant impact on the configuration.

13

Networks of the Future

Networks will interconnect, interrelate, and synthesize various aspects of telecommunication, including many on the horizon. Consequently, the more we know about these emerging subjects, the easier it will be to incorporate them within the network. We study these subjects not only for understanding but in order to visualize them operating within the network.

13.1 Future Configuration

How will the telecommunication network of the future look? Figure 13.1 shows the various interfaces and devices that the network must be capable of handling. This configuration shows the network from a switch standpoint. The facilities for interconnecting these switches must be considered along with the bandwidth needs.

13.2 Network Nodes

The present plans, features, and services are sure to occupy 120% of the available time without the need to postulate the future. However, there always is some value to attempting a view of the future. For example, there is every reason to believe that common-channel signaling and fiber-optic systems yielding tremendous bandwidth will be evident. Basic switching systems can be designed with a few chips and placed at or near the customer's instrument.

**Fig. 13.1. Today's
Communication
Network**

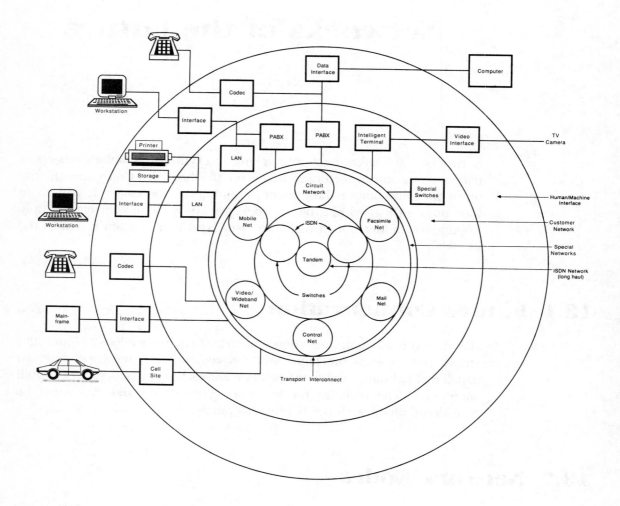

Based on this, it is evident that the instrument will be taking a lead role in the evolution of telecommunication. The network will change as different nodes for different functions are introduced. This will occur because the development cycle is shorter for functional nodes, and in a competitive environment, development cycles and costs are critical factors. The various nodes within the network are expected to be

- A switching node for customer interface and basic calls
- A routing node for determining special or cost-effective handling
- A feature node where common information related to special customer's services can reside

Switching Node

The switching node is obviously the most important part of the network architecture because it is the interface device between the customer and the network. The customer could request voice service, operator service, data service, data-base access service, video service, or any other service available in the near future. The customer also could request centrex service with its common set of features and services. In this case the switching node must connect the customer with a feature node containing centrex features.

One advantage of having a switching node is a shorter development cycle for these switches, thereby introducing new accesses and ISDN arrangements quicker. This means new switching services can be offered without the customer waiting years for the operating company to provide them or going to a bypass company for the service.

Routing Node

The routing node must respond to the various international standards, X.25 or SS#7, to interface the local and long-distance offices within the network. It must be capable of economical route selection by time of day and bandwidth for private networks.

The routing node will have the capability of combining or multiplexing slow-speed lines into higher speed and working out the economics for the calls. The routing node will interface with the switching and feature nodes.

Feature Node

The feature node will be telecommunication's answer to the data base. Some examples of feature nodes would be local video test, library information, films, and stocks. This area is where the entrepreneur will be able to operate by offering special features or services other vendors would find unattractive. It will be the area people associated with telecommunication will discuss and debate quite emotionally.

This is a jaundiced view of the network of the future, but one that appears to be emerging as technology changes our views of switching and transmission.

13.3 Network Services

The present telecommunications provide services in today's environment that can be defined as switch services. That is, the call forwarding, abbreviated calling, and others are provided by the individual switch. The new environment brought about by ISDN and a host of private networks will initiate a demand for new and unique services that networks can offer. Some are

Universal network service Provides users with the ability to define their own network strategy, including numbering plan, service criteria, and routing.

Inward dialing service Allows users to dial into their network from other networks and control the call flow via identification and special routing instructions.

Desk-to-desk calling Allows users to dial anywhere on the network via a specific numbering plan, regardless of whether units are equipped with direct inward dialing (DID).

Centralized management The controls of overloads, switch outages, bandwidth utilization, users' restrictions, billing format, and carrier selection operated from one location.

Screening service Provides users with the ability to restrict access to the network or special computer resources while allowing other calling or resource utilization.

Mail service Routes calls to centralize a voice-mail service with an indication (a light) to the called party that a message is waiting.

Electronic mail will operate the same for individuals equipped with a terminal. Some mail services will provide an option for forwarding caller identification to the storage unit.

Devices will be available to allow networks to integrate the switches and offer the services mentioned here and others which haven't been dreamed of yet.

Telecommunication will provide the incorporation of a multipurpose network to serve an entire organization's communication needs from research and development to sales, and from low-speed data to videoconferencing. It will be capable of serving offices or homes (electronic cottages) within a community or a multinational corporation. With ISDN as part of the network it can be based on capability rather than compatibility and utilize the communication media best suited for a particular requirement. It will have access, both on-net and off-net, to a vast reservoir of information.

13.4 User Control

A major change in both the network and in the customer set will be the movement of the network control from the switch (or switches) where it presently resides to the user. A poor interpretation of this would be the abandonment of controls from the switches, although this is not the intention. There is no intention of changing the very foundation of the industry but merely to instill within the users the ability to control the call as it progresses through the network.

Presently, the user passes the destination code to the central office, and this switch or another will have control of that call until the destination is reached. The builders of these switches are almost incapable of thinking of a network controlled by the user: therefore, we can deduce a reluctance on their part to implement such a radical change. No matter what their argument is or what their vantage point may be, the change is coming! It follows that the experimenters will look for ways (mostly software) to gain control while the network implementers state categorically that it can't be done. But it will be done.

The obvious example from history is the introduction of step-by-step equipment that removes the control of the network from the operator and, for the first time, places this control in the hands of the customers. Bell's great thinkers predicted chaos among other happenings. Only because of an operator's strike forcing the introduction of this equipment did Bell finally accept this innovation. Today, the features of call forwarding, call diversion, and others are basically controlled by the switches. These controls, over the next few years, will evolve to the customer to permit choices—a normal call, a forwarded call, a voice-mail call, a diverted call, or other special handlings. And these choices can be changed during the actual progress of the calls. In other words, if a busy or unanswered line is encountered, the customer will have options available other than "hang up."

How It Will Work

Both the originating and terminating ends of the connection will need equipment to implement this customer control. Therefore the private networks initially will probably be the first to use this approach. Either a device at each end of the network permitting a "handshake" during a call setup to divert the call and to receive instruction on the new destination for the call or an auxiliary unit with access to data bases at both ends permitting the same reconfiguration will be used.

An argument from the present network is that this reconfiguration is not possible because the network's transmission plan will not permit a call from New York to California to divert to Texas. The present analog network has this restriction, and if you ever made a three-way call involving three sectors of the country you will understand their argument. A digital network should have no such restriction; it should have the ability to add gain for the third leg, if required. Perhaps the whole concept of networking must be reexamined. Vendors will offer advice on how to optimize the network with regard to the placement of switches. Other vendors offer the optimal approach to the replacement of the analog switches with digital switches. No vendor, to my knowledge, offers an optimal configuration for the customer's use with minimum cost and flexibility.

Characteristics of a Digital Subset

The customer terminals will incorporate many functions, the actual number will depend on acceptance and economics. It is expected that a gradual transition toward full service will occur, and this period could last a couple of decades. It is apparent that a modular approach is necessary to cater to this long period and the system must be flexible as these services change.

A function that should be obvious by now is the need for the interface between the subset and the connecting office to be digital. The placing of the analog-to-digital conversion at the subset makes service offerings easier but does necessitate the synchronization of the line with the office.

Another essential function will be the four-wire/two-wire conversion at the subset if the loop remains two-wire to the office. The incorporation of these two functions into the subset will allow the customer to have several services simultaneously.

Once these functions are ensured within the subset, the transmission and protocol between the subset and the network will be the critical areas. Meeting the overall transmission characteristics in an all-digital environment should be straightforward and not present many problems. This is not to say that the transmission requirements should be ignored, but it is only to place them in comparison to the combination analog/digital network. The latter will require precise standards that must be checked periodically to ensure compliance.

Transmission standards are being developed by the CCITT subcommittees, and copies of these standards will be available to individuals and organizations. It is highly recommended that copies of these standards be obtained as a guideline to any system being purchased.

When PBXs are inserted within the network, they should be transparent to the connection between the subset and the office; that is, they should contain no loss. With digital four-wire PBXs, this should not present a problem.

Protocol standards basically fall into the same category, as subcommittees strive to define international protocols for the vendors and users.

The quality of speech transmission is going to be improved as the industry moves to an all-digital environment, and this in itself is a noble goal. However, the industry is striving not only for better speech transmission but also to permit additional services. These new services will require an open network that permits various interconnections and configurations. The mere fact that the network has moved from the constraints of an analog environment is sufficient reason for rejoicing and the offering of new services.

13.5 Artificial Intelligence

Artificial intelligence (AI) has for some time been hyped as the next generation of computers and telecommunication. It has been touted by

Japan as the point at which they will take world leadership in the electronic industry. The U.S. electronic industries have put together groups to assist in developments leading to AI to answer the challenge. With this level of activities there is every reason to believe that AI will be integrated with the present information age.

The drive toward AI is caused by the partial success of robotics and the need for a more natural interface to computers. Although some progress has been made with the introduction of application-oriented program (i.e., Lotus 1-2-3), the world of computers is still the domain of the programmers and the specialists who operate the centers and understand the lexicon. It is believed that the introduction of AI will bring the computer and its support networks to the average individual. Artificial intelligence can be categorized as follows:

1. Natural language processing gives users the ability to address data in a data base in a manner that is totally natural. That is, no rules, syntax, or specific processing orders are required to be learned. The system will be capable of detecting ambiguities and prompt the user for the proper meaning. It is constructed in a manner similar to the basic information flow with the computer or equivalent as the receiver.

2   Expert or knowledge-based systems allow the decisions that are normally made by an expert to now be made by the computer. The rules for the making of these decisions are defined by the expert and programmed into the machine. Rules are added as information about the application increases. Its greatest application appears to be the replacement of semiroutine tasks—that is, areas where good assumptions can be made.

3. Image recognition allows computers to recognize shapes, sizes, and depths of objects. It will be an extension of robotics, and it brings this area into the field of AI. The way we perceive objects is addressed, starting with parts and the shape, color, or distance.

4. Machine learning serves as the method by which machines can improve their performance through experience. They must distinguish between useful experience and bad experience.

Although many people believe that AI will evolve naturally over the next few years because all the items listed are available or in work, there are some shortcomings with today's technology to support this new science. The most prevalent argument for AI is the establishing of "new" standards for the computers, which will be introduced after some consolidation of the present market has been achieved. The introduction of new components for these developments also will place a stake in the ground for the manufacturers.

Expert systems will also become an important asset of corporations as they attempt to improve the research and development cycle through the use of these techniques. However, it also presents an opportunity for some in-fighting as parts of an organization lay claim to the programs and the development of the programs.

A scattering of AI activities has occurred since the computer became popular, the most famous being the chess-playing capability. A current example is the scanning devices giving checkout clerks in supermarkets assistance by optical scanners and voice-response outputs.

Natural language processing systems currently can scan a user's request for key words, provide a feedback for verification, and, if approved, retrieve the information from a data base for display on a screen. Most of these systems presently have their own lexicon, depending on the application for which they were developed. The jargon is normally imbedded in the function of the company or the profession; a prime example would be a data-base retrieval system for the law profession.

Future systems will require a more natural interface, quicker response, and a larger data base. Access to the data base will be over a high-speed data link that will be part of the telecommunication network. The individual will be able to sit at a terminal in the office and have rapid access to the data base. In theory the major change to these systems in the future will be the natural interface, but that will present a challenge in cost and standards.

The expert, or knowledge-based, systems will receive the major emphasis in the next few years as companies look for means to assimilate the data and interrelationships needed to make technical decisions. A typical example is the area of telecommunication and data. To design a PBX, for example, we must know the voice and data standards along with the interfaces to LANs, data networks, long-distance networks, special subscriber lines, and operators. It is extremely difficult for an individual to gather all this information and be aware of the changes going on daily in each of the fields. Although not possible today, the gathering of information from each of these areas into a data base would be extremely valuable to the designer of the switch.

The airline reservation system is an example of how such a system would operate as well as the value to the airline and the customer. Many companies are working on these systems as the technical requirements in many fields become more complicated and the value of employees increases.

Present computers are not capable of storing, sorting, and retrieving the information necessary to be considered an expert in special fields, but the fifth generation could produce such a system. The most natural change will be in the world, as a machine will be capable of duplicating the diagnostic ability of a doctor and make recommendations. It will be interesting to see how the AMA and the law profession deal with this problem. A program of this magnitude could bring medical aid to Third World countries.

There is much activity into AI in the labs, but the need exists for focused attention to particular problems and for applying these activities. This is possibly the next level of attainment for AI.

13.6 Global Telecommunication

The advent of AI into the telecommunication arena will solidify the tie between the computer and the network. From this we will be able to see, listen to, or talk to anyone in the world either directly or via stored information. This consolidation of one-to-one communication will have the emphasis of both the computer and the telecommunication industry during the next five years. The various LSI manufacturers will have incorporated ISDN standards on chips during this period, so the vendors within these industries can produce the interfaces needed at a reasonable cost.

About 1991 or 1992, when the network is firmly established with worldwide communication from workstations within the home, we will start to see a merging of the broadcast industry with this new telecommunication. Before this time I would expect both industries to be functioning to stay competitive and current with technologies. By the 1990s videoconferencing will be firmly established with one or two codecs providing full-motion video. It will be considered a viable alternative to travel and a better way to present information. Since videoconferencing transcends both industries, it will be the natural battleground for one-to-one communication and broadcasting. The competition between the industries will require another five years, but the result will be total communication from the user's premise for any type of entertainment media, voice/video communication, or information gathering to any place in the world.

The next 10 years, if the foregoing assumptions are true, are going to be the most exciting era for telecommunication, after which this industry is going to be the dominant industry in the world, displacing the automotive business as people become less dependent on one form of transportation and more dependent on total communication.

When I try to forecast which company will dominate this new telecommunication, I'm reminded of a particular horse race. In this race there were 9 or 10 horses but only 2 were true runners; the rest were "dogs." I proceeded to divide my last $50 between them (it had been a bad day and this was almost the last race) in a mathematically even manner. The race started, and the two horses broke quickly away from the rest of the pack. Going around the last turn for the stretch, they were running neck and neck, but at least 20 lengths in front of any other entry. I was counting my money. As they started down the stretch, the first horse fell, and the second horse tripped over him, allowing the rest of the pack to catch and pass these two favorites. To this day, I don't know who won the race, because I was on the way out by the time it ended. The point is, there are no sure things in horse races or telecommunications. I will predict, however, that the chip manufacturers will be the strongest force in the industry as more and more complex standards are placed on these devices.

Epilogue

We have now reached the end of our expedition through telecommunication networks. I hope that the enormous potential for this effort and its impact on our waking and working is evident.

The purpose of the book was to help develop perspicuous individuals who can guide us in the next 10 years through the labyrinth of changes and innovations. Another objective was to assist the individuals given the task of building a network, and they can use an abundance of support.

I hope to be providing updates to telecommunication technology. If you want to receive these bulletins, please let me know. I also would appreciate any comments on the book.

Glossary

Access system A system where a user accesses a communication network, including customer premise and interface devices at the central office.

ACD, automated call distribution Incoming calls are uniformly distributed over answering positions.

ADCCP, advance data communication control procedure A bit-oriented control procedure for information on a communication network.

AI, artificial intelligence The incorporation of humanlike functions into the next generation of computers. These functions include natural language processing, expert systems, image recognition, and machine learning.

Alternate routing An alternative communication path used if the normal one is not available. More than one route may exist for a particular call.

Analog transmission Transmission of a continuously variable signal as opposed to a discretely variable signal.

ANSI, American National Standards Institute A voting member of the ISO for standards on communication.

Area code See NPA.

ARPA, Advanced Research Projects Agency Contracted for one of the first data networks that used packet switching, the Arpanet.

ASCII, American Standard Code for Information Interchange An eight-level code for data transfer adopted by the American Standards Association to achieve compatibility between data devices.

Asynchronous transmission Transmission in which the individual characters or words are individually synchronized by start and stop elements.

Attenuation The difference between the transmitted and received power due to loss through equipment, lines, or other transmission devices.

Audio conferencing A conference in which people at different points can hear/speak with everyone.

Audio frequencies Frequencies that can be heard by the human ear (usually 30 to 20,000 Hz).

Automated directory An on-line directory for resolving user's identification that can be accessed by anyone on the network via a terminal.

Average busy hour The time-consistent hour within a working day for a period of several days in which the traffic loads are the highest. Sometimes called busy hour.

B channel A 64-kb/s channel within an ISDN interface suitable for voice or circuit-switched data.

Bandwidth The range of frequencies available for signaling. The difference expressed in cycles per second (hertz) between the highest and lowest frequencies of a band.

Baseband frequency The original signal from which a transmission waveform may be produced by modulation.

Baud Unit of signaling speed. The baud speed is the number of discrete conditions or signal events per second. Named for the French inventor J. M. E. Baudot.

Binary code An electrical representation of quantities expressed in base 2.

Bit Contraction of *binary digit*, it is the smallest unit of information in a binary system. A bit represents the choice between a Mark or Space (1 or 0) condition.

Blocking Inability to connect two idle lines (links) because no circuits are available or because no idle connections exist to idle circuits.

Buffering A technique used in data transmission systems to balance traffic to the capacity of some part of the system.

Bursty information Information that flows in short bursts with relatively long silent intervals between.

Bus network A bus structure for LANs with multiple devices all connected to a single line that runs the length of the network.

Busy hour See average busy hour.

Bypass Methods of establishing telecommunication service without employing the telephone company.

Capacity The traffic usage a circuit group can carry or the number of attempts the processors associated with the switch can handle within a prescribed grade of service. Also, the number of terminations the switch can physically accommodate.

CAROT, centralized automatic reporting on trunks Performs end-to-end testing of the various trunks (links) in the system.

Carrier system A means of obtaining a number of channels over a single part by modulating each channel on a different carrier frequency and demodulating at the receiving point to restore the signals to their original form.

CCIS, common-channel interoffice signaling A special purpose network used by switching systems to communicate with each other.

CCITT, International Telegraph and Telephone Consultative Committee A group for international agreement on recommendations for international communication systems.

CCITT standards Interface and protocol definitions recommended by the CCITT.

CCS, centum call second A measurement of traffic usage common in North America. It means 100 call seconds of occupancy during the study period. One CCS is equal to 1/36 of an erlang.

Centralized attendant The ability of switching equipment to forward attendant calls to one location regardless of where they originated.

Centrex A switching unit that operates similar to a PBX but allows every subscriber to be directly dialed from the outside.

Circuit The physical connection of channels, conductors, and equipment to provide discrete communication between two given points.

Circuit switching A method of communicating where a connection between calling and called stations is established on demand for exclusive use until the connection is released.

CMR, cellular mobile radio A system that allows voice and control transmission from a mobile station (car) to a stationary ground control center.

Coaxial cable A widespread and moderately priced cable with large capacity and low error rates.

Codec Coder-decoder used to convert analog input into digital format and digital input into analog format.

Common carrier A private or public corporation responsible for providing telecommunication service in a given territory.

Compressed video Television signals transmitted with much less than the usual bit rate. Full standard coding of broadcast quality typically requires 45 to 90 Mb/s. Compressed video is in the range of 64 kb/s.

Concentration In telephone systems, a switching network (or portion of one) that has more inputs than outputs.

Concentrator A communications device that combines many low-speed or low-occupancy lines onto one or more high-speed or high-occupancy lines. In data, functions are similar to a multiplexer. In voice, refers to a suboffice connected to a computer-controlled switch.

Controls The equipment shared within a switching system only during the periods of a call needed to accomplish a function or functions.

Crossbar switch A switch having a plurality of vertical and horizontal paths and using electromagnetically operated devices to connect one vertical path with any horizontal path.

CSDC, circuit-switched digital capability A 56-kb/s alternative voice/ data circuit-switched feature.

D channel A 16-kb/s channel within an ISDN interface suitable for packet data and user-to-network signaling.

dB, decibel A tenth of a bel. A unit for measuring relative strength of a signal parameter such as power or voltage. The number of decibels is 10 times the logarithm (base 10) of the ratio of the measured quantity to the reference level. The reference level, such as 1 milliwatt for power ratio, must always be indicated.

DCE, data communication equipment The equipment that provides the functions to establish, maintain, and terminate a connection. The interface between the data terminal equipment and the data circuit.

Dems, digital electronic message service A T1 service for various opportunities, such as data, telemail, facsimile.

Dial pulse A current interruption of the DC loop of a telephone. It is produced by the breaking and making of contacts when the digit is dialed and the rotary unit is returning to normal.

Digital signal A discrete or continuous signal; one whose various states are discrete intervals apart.

Digital switching A method of switching voice or data traffic by encoding them in a digital signal format before passing the calls through the system.

DIOD, direct in/out dialing The ability of a PBX or equivalent switch to receive and send calls to the public without the intervention of an operator.

Directory See automated directory.

DLP, decode level point Level of the analog signal after being decoded from a digital format.

DTE, data terminal equipment The equipment providing the data source or sink. Normally connects to data communication equipment.

DTMF, dual-tone multifrequency A method of signaling in which two frequencies are combined, one high and one low, to indicate a digit value to the switching system.

DUV, data under voice The passing of data information over a circuit occupied by a voice connection. The data is normally passed during different frequencies in the connection from the voice connection.

EBCDIC, extended binary coded decimal interchange code An 8-bit character code used primarily in IBM equipment.

EIA, Electronics Industries Association An association for providing standards for computer/telecommunication vendors. RS-232 is an example.

Electromagnetic waves Both radio and light are examples of electromagnetic waves, which can be propagated by a transmitter and broadcast through the atmosphere.

Electronic mail Mail forwarded over communication channels and received via a computer or terminal.

ELP, encode level point The level of the analog signal before being encoded to a digital format.

Erlang A dimensionless unit of traffic intensity. One server, fully occupied for the study period, equates to one erlang. Also, the load in erlangs is the average number of servers occupied during the study period.

Expansion   In telephone systems, a switching network (or portion of one) that has more outputs than inputs.

Expert systems   Allows the decisions normally made by experts to be made by the computer. Also known as knowledge-based systems.

FACS, facsimile   A system for the transmission of images. The image is scanned at the transmitter, reconstructed at the receiving station, and duplicated on some form of paper.

FDM, frequency-division multiplexing   A multiplex system in which the available transmission frequency range is divided into narrower bands, each used for a separate channel.

FEP, front-end processor   A node with network control program to interface to mainframe or host processor.

Full-duplex   Simultaneous two-way independent transmission in both directions.

FX, foreign exchange   A feature that allows a local number to be part of one office while the actual connection occurs in another office.

Geosynchronous satellite   A satellite in orbit 22,300 miles above the equator and traveling at a speed where it appears to be stationary relative to a point on the equator.

Half-duplex   A circuit designed for transmission in either direction but not both directions simultaneously.

HDLC, high-level data-link control   A bit-oriented control procedure established by the ISO.

Hertz   A unit of frequency equal to cycles per second.

Hierarchical network   A means of routing calls in a network where a call is passed through offices of different classes toward its destination.

High day   The highest business day busy-hour traffic. Processors, markers, special links are often sized from this criteria instead of average busy hour.

Holding time   The length of time a circuit or facility is held in use to carry and/or process a call.

IEC, interexchange carrier   A common carrier for handling traffic between LATAs.

IEEE, The Institute of Electrical and Electronics Engineers A professional organization that supplies standards in addition to information to its members.

Image recognition Computers will be capable of recognizing shapes, sizes, and depths of objects.

Interconnect To connect privately owned components to a public network of communication.

ISDN, integrated services digital network A standard approach to digital networks providing integrated access for a multiplicity of user services.

ISO, International Standards Organization Issues recommendations for the design of communication networks.

ITU, International Telecommunication Union The telecommunication agency of the United Nations, established to provide standardized communications procedures and practices including frequency allocation and radio regulations on a worldwide basis.

KTS, key telephone system A key-selected access of a customer station to a multiplicity of links, offices, or other lines.

LADT, local area data transport Used by data-switching vendors for slow-speed terminals that commonly use data-over-voice transmission scheme. Data over voice uses the silent period of a voice connection to transmit data.

LAN, local area network A network of data devices and computers not connected to a national network.

LATA, local access and transport area An area where a phone company provides local or intraLATA service and access to IECs for other calls.

Lattice structure A gridlike structure for routing calls through a network.

Link A circuit and/or transmission medium connecting two points, offices, exchanges, or nodes for either a voice or data call. Also called circuit, trunk, trunk circuit, virtual circuit.

Machine learning Machines will be programmed to improve their performance through experience.

MAP, manufacturing automation protocol A token-passing bus designed for factory environments by General Motors.

MDR, message detail recording The billing information of a call, including, at least, the calling and called numbers and the call duration.

MERS, most economical routing scheme A method where the least expensive group to the most expensive group is followed.

Microwave Any electromagnetic wave in the radio frequency spectrum above 890 MHz.

Modem A contraction of modulator-demodulator. The term may be used when the modulator and demodulator are associated in the same signal-conversion equipment.

Multiplexer A device for connecting several devices at the same time to communication lines or computers to control the transmission and reception of calls.

Multiplexing The division of a transmission facility into two or more channels by splitting the frequency band transmitted by the channel into narrower bands, each of which is used to constitute a distinct channel (frequency-division multiplexing), or by allotting this common channel to several different information channels one at a time (time-division multiplexing).

Murphy's law (1) An optimistic view of system development. (2) A scheme whereby previous definitions are used in the glossary to avoid NIH (not invented here) factor.

Natural language processing The users will be able to address data in a data base in a totally natural manner. No rules, syntax, or specific processing orders will be required.

NBS, National Bureau of Standards A division of the U.S. Department of Commerce, which is funded by the federal government.

Network access Normally refers to the equipment or protocol necessary to interface to a particular network.

Network control center A facility for monitoring and controlling the links and nodes associated with the network. Is also used to pass the administrative data of the nodes.

Node An end point of a link or a junction common to two or more links in a network. In voice communication, normally referred to as a switching center; in data, a primary or secondary center.

Nonhierarchical routing A flexible means of routing calls in a network that allows any node to act as a tandem-switching point.

NPA, numbering plan area A three-digit number identifying one of the geographical areas of the United States and Canada to permit direct dialing on the network.

OPX, off-premise extension A line connected to a switching system (PBX) given special treatment due to the distance involved.

Overlay network A separate network for a particular service or strategy covering most of the same geographical locations as the basic network.

Overload management Handling peak demands by selectively delaying, degrading, or dropping certain calls and completing only those portions of traffic flow that will yield the highest completion rate.

Packet A group of bits including data and control procedures that is switched as a whole. The bits are arranged in a specified format.

Packet switching A transmission process, using addressed packets, whereby a channel is occupied for the duration of the transmission. Commonly used in data transmission systems.

PAD, packet assembly/disassembly The interface between the user's information and the format for transmitting information on a data network.

Parallel transmission Simultaneous transmission of bits over separate channels making up a character or byte (8 bits over 8 lines).

Parity check Addition of noninformation bits to data, making the number of 1's in each grouping of bits either always even or always odd. Permits single error detection.

PBX, private branch exchange An exchange connected to the public network on the user's premise and operated by the user. Normally, outgoing calls are automated dialed, whereas incoming calls are handled by an attendant.

PCM, pulse code modulation Modulation of a pulse train in accordance with a code.

Peakedness A measure of the nonrandomness of traffic on a circuit or group of circuits.

Picturephone A trademark name of the Bell System for video communication. Normally refers to the terminal in front of the user.

Ping-pong A data transmission scheme where the portions of the call, incoming and outgoing, are alternately switched onto the channel.

Polling A method of controlling or collecting information from a communication line.

Protocol   A formal set of conventions governing the format and relative timing of message exchange between two communicating processes.

Quantizing or quantization   Process associated with PCM involving the division of an analog signal into digital form by sampling and the reassembling of the digital signal after transmission into an analog waveform.

Real time   A real-time system is controlled by certain service criteria within an environment, such as receiving information, processing information, switching times, and disconnect times.

Recent change   A method whereby the switching system is updated on the latest moves and changes within the system.

Redundancy   The equipment supplied to ensure that the flow of information is not interrupted if loss or failure is encountered.

Repeater   A device used to restore signals that have been distorted because of attenuation to their original shape and transmission level.

Ring network   A network, common in LANS, where each device is connected to adjacent devices.

ROTL, remote office test line   Test equipment located within the office where the trunks (links) are to be tested. Used with CAROT.

Routing   The assignment of the communication path by which a message or telephone call will reach its destination.

R-series   A set of protocols to assist the current design to interface to ISDN.

Rusty switch   Normally a line hung over a state boundary to take advantage of interstate rates on certain calls.

Satellite   An orbital space station housing on-board power, orbital controls, and transponder (repeaters) for receiving and relaying messages over a wide range.

SDLC, synchronous data-link control   A bit-oriented control procedure introduced by IBM in 1969.

Serial transmission   A method of transmission in which each bit of information is sent serially in time on a single channel.

Service criteria   Either lost call cleared (busy tone) or lost call held (dial tone delay) at a specified grade of service.

Signal-to-noise ratio   Relative power of the signal to the noise in the channel.

Simplex mode Operation of a communication channel in one direction only, with no capability for reversing.

SNA, systems network architecture Description of structure and operational sequences for transmitting information through and controlling configuration of networks.

Software A set of programs, procedures, rules, and associated documentation concerned with the operation of network computers and processors.

Space-division switching A switching system that has a physical connection for each conversation within the office. May be electronic or electromechanical. Controlled by a processor.

SPC, stored program control A telephone switching system controlled by one or more electron processors that direct all operations. System may be analog or digital.

Special-purpose network A network designed for a single purpose and specializing in that, as opposed to a general-purpose network.

Spectrum A continuous range of frequencies, usually extensive, within which waves have some specific common characteristic.

Star network A network commonly used for LANs; a PBX is an example.

Start/stop system A system in which a group of code elements corresponding to a character signal is preceded by a start signal to prepare the receiving unit and is followed by a stop signal to indicate the end of that transmission.

Subscriber's line The telephone line connecting the exchange to the subscriber's station.

Switchboard The unit manned by the attendant for handling calls to and from the affiliated system.

Switched traffic A method where traffic, although both originating and terminating, is only counted once as opposed to counting it at the originating and terminating sides.

Switching system A system used to switch telephone traffic in the network. Can refer to an exchange, an office, or SPC machine.

Tandem exchange An office used to interconnect local end offices over tandem links in a densely settled exchange area where it is uneconomical for a telephone company to provide direct interconnection between all end offices. The tandem office completes all calls between the end offices but is not directly connected to subscribers.

TASI, time assignment speech interpolation A technique that only assigns a channel to a conversation when speech is occurring.

TDM, time-division multiplexing A system of multiplexing in which channels are established by connecting terminals one at a time at regular intervals by an automatic distribution.

Teleconference An audio conference or a videoconference.

Telemail See electronic mail.

Terminal Any device capable of sending and/or receiving information over a communication channel.

Tie line A private-line communication channel of the type provided by communication common carriers for linking two or more points together.

Time-division switching systems A switching system used to create networks for digital traffic in which the techniques of TDM are used.

Token-passing network A LAN network where a token is passed from device to device for communicating.

Toll office Basic toll switching entity; a central office where channels and toll message circuits terminate. A Class 4 office.

Transceiver A terminal that can transmit and receive traffic.

Translator A device that converts information from one system of representation into equivalent information in another system of representation. In telephony it is the device that converts dialed digits into call-routing information.

Trouble log A means of recording the trouble conditions at a site in order to determine patterns and whether service levels are being maintained.

Trunk circuit See Link.

TSPS, traffic service position system Operator service for local and long-distance calls. Automatic billing on long-distance calls.

Usage A traffic usage measurement item. Usage is the measure of the amount of time that the servers within a group were in use during a period of time. The unit of measurement is normally in erlangs.

Virtual circuit A connection between a source and a sink in a network that may be realized by different circuit configurations during transmission of a message. Common in packet networks.

Voice-grade channel A channel used for transmission of speech, digital or analog data, or facsimile, generally with an audio frequency range of 300–3400 Hz.

VTAM, virtual telecommunication access method A collection of standards and units for gaining access to mainframe computers.

WATS, wide area telephone service A service provided by telephone companies in the United States that permits a customer to use an access line to make calls to telephones in a specific zone in a dial basis for a flat monthly charge. Monthly charges are based on the size of the area to which the calls are placed, not on the number or length of calls.

Workstation A unit incorporating the features of a computer terminal and a telephone.

X . . . series The CCITT has produced a number of standards to establish communications interfaces for users' data terminal equipment and for network interface units or modems known as data circuit terminating equipment.

X.21 A member of the family of X series for connection of terminals to modems.

X.25 A protocol for packet-switching networks that supports the attachment of intelligent terminals as well as communication controllers and host processors.

Index